Living Control Systems III:
The Fact of Control

William T. Powers

Benchmark Publications Inc., Bloomfield, New Jersey 07003

Library of Congress Control Number: 86-62634

Powers, William T. (William Treval), 1926-

Living Control Systems III : the fact of control / William T. Powers —1st ed. includes mathematical Appendix, Index, and over 50 illustrations (more than half in full color); Companion CD of interactive models and simulations of Perceptual Control Theory

ISBN 978-0-9647121-8-8 (acid-free paper)

1. Perceptual Control Theory (PCT) 2. Psychology 3. Neuroscience 4. Sociology

Author photograph: Mary A. Powers

Cover design: Williams & Company

LCS III demos: Dr. Bruce Abbott

program updates / interface design

Mathematical Appendix: Dr. Richard Kennaway

Printed in the United States of America

1 2 3 4 5 6 7 8 9 10

Dedication

To Mary Andrews Powers

November 7, 1930 - October 8, 2004

My kindest and wisest critic for 48 years

Acknowledgments

Dr. Bruce Abbott, Indiana University-Purdue University Fort Wayne, who has devoted much time over 20 years to my efforts, showed me how to use Delphi, came up with useful tricks to allow me to avoid learning object-oriented programming, and provided a knowledgeable link with the whole field of psychology.

Dr. Richard Kennaway, School of Computing Sciences, University of East Anglia, UK, who has patiently helped me through problems beyond my mathematical capacities.

Professor Yan Zexian and *Professor Fan Dongping*, South China Normal University, and *Professor Zhang Huaxia*, Zhongshan University, whose generous support and acceptance of PCT have turned into friendship as well.

Dr. Richard Marken, one of the first to understand my modeling efforts, and author of many ingenious and instructive computer models produced in the 30 years since we first met.

Dr. W. Thomas Bourbon, author of the first PCT models of social interactions and a staunch ally who has, since the 1970s, sacrificed much in order to continue his work with PCT.

Dr. Timothy Carey, University of Canberra, Australia and *Professor Warren Mansell*, Manchester University, UK, whose energetic work in teaching has furthered PCT and its clinical derivative, the Method of Levels.

Dr. David M. Goldstein, the first clinical psychologist after Richard Robertson to try the Method of Levels in the U.S.

Dr. Richard Robertson, psychotherapist, who has known of and devoted his efforts to furthering PCT and its predecessors since 1957.

Dag Forssell, current archivist and much more to the Control Systems Group, whose web pages and video recordings contain all the most important events in PCT's history.

Gregory Williams, first archivist of the Control Systems Group and editor-publisher of three of the books about PCT, whose careful readings and mathematical contributions made possible some of the early behavioral models.

Modeling and Simulation Demonstrations CD

SYSTEM MINIMUM REQUIREMENTS

Hardware:	Personal computer
	Speed of 900 MHz or faster recommended
CD-ROM drive	Any standard CD-ROM drive
Memory:	One Gigabyte recommended
Screen resolution:	800 x 600 or better recommended
Hard drive space:	100 Megabytes of disc space recommended
Operating systems:	Tested on Windows XP (Home, Media Edition, and Pro), and Windows Vista (not the minimum versions).
	Tested on Windows XP under VMWare Fusion and under Parallels (two environments for running virtual Windows machines) on Apple Macs. The tester suggests the programs might also work under other virtualizers available on the Web, some free. Newer Macs, which use Intel processors, can be booted into Windows with the Leopard operating system or BootCamp (part of Leopard).

CD INSTALLATION

In Chapters 2 through 9, the discussion is supported by one or more interactive computer programs to give you hands-on experience with perceptual control theory principles.

The CD is set up menu style. Load the CD and open it. Then double-click LCS3Install.bat. The demonstration programs will be copied to the C drive in a new folder called LCS3Programs. The red ball icon (LCS3Menu) appears on the desktop. When you double-click it, a menu will appear on the screen. Highlight the name of the demo called for in the book. Start the demo by clicking Run Selected Demo.

To facilitate your transition from reading concepts to trying the demonstrations, as you call up each demo to your screen, take a minute to review the control panels and display windows. Then follow the steps (bulleted text) in the book. You can also click the on-line instructions button, available on most screens, for tips and reminders.

To avoid clutter, many popup windows close only by using the X in the upper-right corner.

NOTATION

CONTROLS	TO ACTIVATE
LABELED BUTTONS	single click
Radio Buttons	click the new selection
Sliders	click-drag the pointer
Spinners	click the up-down triangles
Min/Max/Close box	click x to close the program

William T. Powers

Author of LCS III and originator of PCT and the models and simulations

Bill Powers grew up in a suburb of Chicago, Illinois and served in the United States Navy in World War II, after which he earned his Physics degree from Northwestern University on the GI Bill. His interest in control theory as a model of behavior began while he was a medical physicist at the Veteran's Administration Research Hospital in Chicago, where he designed many devices for medical research; notably, a curve-tracer for plotting isodose contours in the beam of radiation from a Cobalt-60 therapy machine.

While attending the Graduate School of Psychology at Northwestern in 1960, he worked part-time as a technician in the Department of Astronomy where later he became chief systems engineer. In that capacity, Powers designed and built low-light-level television systems for astronomy. He was involved in the design of the Lindheimer Astronomical Research Center, and was responsible for the design and building of the Corralitos Observatory. He created the automatic all-sky photometer for use on the moon (for Apollo 18, which never flew).

During this period, Powers conducted several one-quarter seminars on control theory at NU, sponsored by Don Campbell and Hugh Petrie. Throughout these early years, he was publishing regularly on the subject of control theory (his book, *Behavior: The Control of Perception*, was the culmination of 20 years' study when it was first published in 1973).

In 1984, Powers and his colleagues founded the Control Systems Group (CSG); members participate in research, discussion, and practical application of control theory, now called perceptual control theory (PCT). They communicate on line regularly, and hold an annual International Conference (see http://www.perceptualcontroltheory.org).

Powers retired in 1990 from the Chicago Sun Times, where he spent over a decade as a technician in the technical services group. Among many special projects, he developed a system for receiving newsprint manifests by wire, and won the Marshall Field Award for his microcomputer system for receiving, formatting, and typesetting satellite-broadcast stock tables in real time. Throughout his varied career, and ever since, he has continued his work on developing a control-system model of behavioral organization. In retirement, Powers spends his time writing, giving seminars and papers at scientific meetings, conducting research and working with colleagues on collaborative projects, such as building testable models of the theory. When asked to describe himself, Bill said:

"I've always been pretty much a blue-collar type,
by training and by preference."

Contributors: Brief Biographies

The two contributing experts to *Living Control Systems III: The Fact of Control* are both actively engaged in research, teaching and writing at their respective universities. They have dedicated a great deal of time to the study and development of perceptual control theory, in addition to their other responsibilities and areas of expertise.

Dr. Bruce Abbot redesigned, updated and improved the demonstrations offered on the CD that accompanies this book.

Bruce Abbott graduated from Bowling Green State University in 1980 with a Ph.D. in Experimental Psychology and has been a faculty member of the Department of Psychology at Indiana University Purdue University Fort Wayne since then. He has published research on animal learning and behavior and is a coauthor of Research Design and Methods: A Process Approach (McGraw-Hill). He is currently exploring the application of control theory within the field of learning and behavior.

Dr. Richard Kennaway authored the Appendix for LCS III, which describes in illustrated detail the mathematics behind the scenes in the 13 interactive demonstrations of PCT at work.

Richard Kennaway studied mathematics at Edinburgh and Oxford, receiving the D.Phil. in 1981. Since then he has been engaged in research in computer science at the School of Computing Sciences, University of East Anglia, UK, in the areas of programming language theory and design, virtual reality, control theory, robotics, and biocomputing.

Contents

Figures

Foreword

This work began in the second sub-basement of the Argonne Cancer Research Hospital at the University of Chicago in 1952, when I was all of 26 years old. That was the year I gave up my interest in dianetics and went back to physics, the same year my friend Kirk Sattley showed me a small book called *Cybernetics: Control and communication in the animal and the machine*, and unwisely let me borrow it. Kirk died in 2002, but I still have his book. In my new job as a very junior health physicist, I had time to do some experiments with human control systems. One of them was an attempt to use external positive feedback to measure the amount of internal negative feedback in the galvanic skin response to loud sounds. There is no need to inform me that this experiment was rather confused in its premises. So was I. But I did begin to study engineering texts on control theory, and the work started that led, over many years, to this book and four others before it.

Now, with this book, I seem to have done about as much as I can to preserve my version of this thread of thinking and pass it on to those who will make the best use of it. It still seems to me, as it seemed an hour after putting down Norbert Wiener's book 55 years ago, that negative feedback control is the ingredient that has been missing from the life sciences for as long as they have existed, and is the concept that will finally put those sciences on a par with physics and chemistry. I can't begin to guess why it has taken so long, why so many people have explored the phenomena of negative feedback control in living systems without seeing that they were looking at a revolution straining to happen, and why so many more have simply refused to learn anything about it. But that's their problem.

In this book I have only one goal: to establish in the mind of the reader the literal reality of negative feedback control as the basic organizing principle of human behavior. Human beings do not plan actions and then carry them out; they do not respond to stimuli according to the way they have been reinforced. They control. They never produce any behavior except for the purpose of making what they are experiencing become more like what they intend or want to experience, and then keeping it that way even in a changing world. If they plan, they plan perceptions, not actions. If they respond to stimuli, they do so in order to prevent those stimuli from affecting variables they have under control. The root, the core, of the behavior of living systems is negative feedback control, at every level of organization from RNA and DNA to a spinal reflex to a mental concept of physics. Negative feedback control is the basic principle of life.

By the time you finish this book you may be more willing than you are right now to join me out on that limb, which you may find is sturdier than it looks from here.

The theme of this book is embodied in thirteen computer demonstrations. Each demonstration shows you something about how living control systems, or negative feedback control systems in general, work. In a number of cases, those control systems are inside of you and you can observe them working. In all cases, the phenomena are perfectly clear and reproducible by anyone; the demonstrations always work, the experiments always produce the predicted results. These are just a few of several important contrasts between the approach we will take here and the usual kinds of behavioral experiments found in the life sciences.

This is my last chance to persuade my peers on this planet that prior to the 1930s we human beings had an entirely wrong idea, only partly supplanted at the time of this writing, of who we are, how we are constructed, and why we do what we do. I hope to do this persuading by showing each person who reads this book a mirror in which certain commonplace phenomena of human behavior can be seen from a new angle. The computer age has provided a tool that allows me to create a small universe which each of you can experience and act upon directly, as if for a time you can ride with me inside my brain, doing what I do, seeing the same results I see, checking my observations against your own, and —if all goes well —reaching the same conclusions I have reached. This is basically a technical book, in the respect that it focuses on computer models and simulations that require some expertise to understand completely. But it is also meant as a teaching book, for anyone who is willing to wade through a few deep places and take a few assertions on trust. I will try to say everything in plain enough language to be understood by anyone, expert or not. The programs can be run and (I hope) enjoyed by anyone who can click a mouse. My colleague in Indiana, Bruce Abbott of Indiana University-Purdue University Fort Wayne, has consented to rewrite all my old demonstrations and experimental programs in a modern computer language and a consistent style, and to contribute a few of his own ideas. My colleague, Richard Kennaway, East Anglia University, UK, has agreed to provide an appendix in which my amateurish approach to mathematics is given some respectability by being done right.

I will do only a little persuading or arguing here. The main goal is simply to make sure as best I can that the points of these experiments and demonstrations are made clear. If you understand them, you will know what I know and you may even agree with my conclusions. I hope you will find things here that surprise and delight and illuminate, as discovering them and working them out did for me over the last half century. To me, this theory is beautiful, not because it is mine but because it is true. That is why I present to you here *the fact of control*, so you can see and feel it for yourself, have that joyful shock of discovery, and understand it.

William T. Powers

Lafayette, Colorado, USA

November, 2008

Chapter 1: Concepts of Control

This book is about control theory and its uses as a way of explaining how human behavior, and perhaps the behavior of all organisms, works. Before getting into the computer demonstrations, we'll spend one chapter on some generalizations that may be helpful, and some claims with which I hope you will eventually agree.

> *I'll make a deal with you. You can skip Chapter 1 if you will agree that before you raise objections about something said in later chapters, you will go back and read it. It's a bit heavy on mathematical concepts and not strictly necessary for understanding what comes next, but I believe the mental exercise will be at least a little helpful. If you disagree, just skip to Chapter 2 and get started with the demonstrations.*

Whatever method of control one proposes, the basic concept of control is the same. To control something is to act to bring it to a specified condition, and then maintain it close to that condition even if unpredictable external forces and changes in environmental properties tend to alter it. Human behavior, as later chapters will suggest through numerous examples, is control behavior. All of it.

Anyone encountering for the first time the brand of control theory used here, PCT (perceptual control theory), might be a little confused if some other kind of control theory has happened to get its pitch in first. PCT is actually based on classical negative feedback control theory, formalized during the 1930s and made into an engineering discipline during World War II, in the 1940s. Classical control theory was the basis of engineering psychology and cybernetics, and is still the primary theory used by most control engineers in designing artificial control systems.

Another kind of control theory has existed for at least as long as classical control theory. We can call it the "analyze-compute-act" model. It has been known under many names: *compensatory responses* in the psychology of the 1930s; modern control theory (MCT) in present time. It's quite possible to invent a way of controlling things that depends only on straightforward (but not simple) logic and physics, if you don't happen to know about classical control theory. That's how I think the other approaches to control got started. They are what you end up with if you reinvent control theory without the classical principles behind PCT.

It will be easiest to understand the differences between these approaches to control if we look at the MCT version first. I'm going to do this in some detail because I want you to appreciate just what is involved in the analyze-compute-act way of controlling. Knowing the differences is important because analyze-compute-act is the version of control theory you will most likely encounter if you read the literature of psychology, neurology, or medicine. That version is a rather formidable obstacle to the acceptance of PCT, so it's worthwhile to take a little time to look for chinks in the stone wall.

MCT: ANALYZE-COMPUTE-ACT CONTROL

The basic idea behind the analyze-compute-act approach is simple and logical; it's based on the way an engineer would analyze a task and build a device to accomplish it. In order to bring about and control some physical effect or result, the engineer would first analyze the situation to see what actions, what forces or other influences, can be used to generate the intended result. For each of those actions, some preceding process would be needed to generate it, with the design not being complete until all required actions can be produced. An input of some kind would then trigger the start of the actions, which would be carried out in a precalculated way. If the analysis was correct and all the effects of the actions were predicted properly and physically generated with the required precision, the desired result would inevitably happen.[1]

That plan sounds perfectly reasonable until you start to apply it to a specific problem. Then you find that the deeper you get into the design the more little difficulties show up, until it becomes plain that designing the system so it can acquire, all by itself, all the data and knowledge it will need, providing the system with the facilities for doing all of the computations fast enough and accurately enough, and building the mechanical and electronic devices that can generate the necessary actions exactly as they must be carried out for the calculated result actually to happen—accomplishing all of that, you would have to conclude, is all but impossible.

That's easy to say as a generalization, so I thought it would be a bit more convincing if I walked the reader through just a sketch of what a design of a simple control system might look like when we explore one or two layers beneath the "simple and logical" summary of the design procedure. If plodding through mathematics doesn't seem very entertaining, just let it roll past and get the flavor of all the little problems that arise in trying to build a system that, all by itself, can analyze the environment, compute how to achieve a prescribed result, and then act on that environment in exactly the calculated way. I'm setting up a straw man here so you will be encouraged to knock it down, but I will play fair and not exaggerate the problems.

Suppose we are designing a control system for controlling the speed of a car on a highway. The MCT approach begins by having the control system itself analyze the physical situation— "the plant," as it is called. This is the *analyze* part of analyze-compute-act. The

1 Sometimes the analysis takes the form of a model of the external world, placed inside the control system.

speed of a car is the result of all the forces that act on it over time, according to Isaac Newton. To change the speed of a car along the road, it is necessary to apply a force to it, using the car's engine, a tow-truck, or gravity if the car is on a downslope. The force will generate an acceleration equal to force divided by the mass of the car. The acceleration will cause the velocity of the car to increase at a rate proportional to the force (the distance of the car from the initial position along the road will increase at a rate proportional to the changing velocity). To compute the velocity V of the car at any time T after the force is applied, the control system can use the equation, $V = V_0 + (F/M)(T - T_0)$. V_0 or "V-sub-zero" as the initial velocity. T_0 is the time at which it first applied the force F to the mass M.

The correct equation is not quite as simple as this. As the car speeds up, air resistance and internal friction in the engine, transmission, axles, and tires produce an increasing force opposed to the engine force, so the net effective force is actually decreasing as the car speeds up. Frictional resistance increases approximately as the square of the velocity, so the control system really has to write

$$V = Vo + [(F - K_A V^2)/M](T - T_0)$$

where K_A is a calibration coefficient.

Unfortunately this is still not exact because now the net force applied to the mass is changing with time, so the control system has to abandon algebra and integrate the nonlinear differential equation, which is

$$dV/dt = (F - K_A V^2)/M,$$

which has a rather complex solution we won't bother with here.[2]

If the road turns uphill, another resistive force will arise which changes as the slope of the road changes. This force will be $-MG[\mathrm{Sin}(s)]$ where s is the upslope angle, G is the (negative, downward) acceleration due to gravity, and M is the mass of the car.

2 Partly because it would take me a day or two to figure out how to solve it.

The angle s is a function of the position of the car along the road, because the slope begins at some position on the road, increases for a few hundred feet until the maximum slope is reached, then decreases again as the top of the hill is approached. The control system has to compute the position of the car along the road as a function of time so it can calculate the number to use for s. This gives a term involving the sine of the slope angle which changes with position. The position of the car is the time-integral of the velocity, and the velocity-changes depend on the position of the car, so the result is a really complex equation with no simple route to a solution. The difficulty is multiplied by the way the slope changes as the car moves along the road; the road does not necessarily follow any simple upward curve. It's necessary to fit some idealized curve to the actual road shape, picking a curve that leads to an equation the control system can—perhaps—solve without causing too much error because of not using the actual shape of the hill.

That's probably enough to give the picture. Simply to compute what the velocity will be at a given time after the car starts moving is far from a trivial task. And now the control system has to make what seems to be a very complex decision. It has to pick what is known as a "trajectory" for the car. It has to decide how it wants the speed to change as the car goes from zero speed to the desired speed. Should the car accelerate rapidly at first, and then gather speed more slowly? Should there be any periods of constant speed before reaching the final speed? Should the car take a long time to reach the final speed, or get there as fast as possible? Whatever trajectory it chooses, achieving it will result not from simply applying a force to the car using the engine, but from *varying* the force, changing it in just the way that will result in the chosen speed-up pattern. The formula for the desired pattern of speed changes has to be substituted for V in the last equation above. Then the system has to try to solve that modified differential equation for the pattern of forces that will result in the desired trajectory. That is the *compute* part of analyze-compute-act. And the control system still has to figure out the *act* part.

Our very intelligent control system has solved the equations stemming from Newton's laws and the laws of aerodynamics, and deduced the pattern of forces that must be applied to achieve the intended behavior of the car. Now it has to figure out how to generate those forces. It must now analyze the car's engine and transmission, the properties of its tires, the quality of the gasoline in the tank, the settings of the engine control computer. From that it can deduce how hard it must press the accelerator down against its spring (having a known spring constant). And then it must produce that pattern of pressures with sufficient accuracy that the entire trajectory is brought into being without unacceptable errors. I will mention without further tedious details that in order to do this, the brain—er, the control system—must know about the properties of its output amplifiers and the current state of the muscles and their attachments to the accelerator. It must also predict the weather, to know whether there will be a headwind or tailwind, and use aerodynamic formulas again to compute the effect of the wind on the speed, if any.

The most important thing to realize is that all the data-taking, all the underlying knowledge about physics, and all the computations have to be done not by a team of physicists or engineers in some laboratory, but by the control system, the driver in the car. These computations are part of the proposed model: the hypothesis about how a brain goes about driving a car.

Certain proponents of the analyze-compute-act model appear to believe that because they can do computations like these slowly on paper, or somewhat less slowly by programming a computer, the brain can do them easily, instantly, and unconsciously. (If you agree with that, you may as well put this book down right now. It's not going to impress you. There is no magic in this book).

Those who are boggling at the idea that accelerating a car to a desired speed while driving up a hill requires this sort of machinery to be acting in the brain will be relieved to know that there is a far simpler way to do the same thing by negative feedback control, the kind used in PCT.

PERCEIVE-COMPARE-ACT CONTROL

Negative feedback control also requires producing forces and other influences as required to generate a desired result. But it accomplishes this end in a way very different from the MCT method. Instead of precalculating what it needs to do and then doing it, the control system starts acting immediately, continuously adjusting its actions on the basis of their ongoing consequences. Negative feedback control as used in PCT can be summed up very quickly. It involves continuously *perceiving* the current state of whatever is to be controlled, continuously *comparing* that perception with the intended state, and continuously *acting* to reduce the difference between perceived and intended states, keeping the difference as close to zero as possible. The feedback part of negative feedback says that effects of the output (action) caused by the difference feed back to the input to change the perceptions and affect the difference. The negative part comes from the fact that the action decreases the difference instead of increasing it. Causation runs around a circle or loop. Such systems work without doing any analysis of the outside world, without "inverse kinematic and dynamic" calculations (the MCT model described earlier), and without even knowing what is causing perturbations of the thing being controlled. Here's how it works when you drive a car.

First, you select the speed at which you want the car to be going—a reading on the speedometer, say 60 miles per hour. This provides a reference condition against which you can compare the perceived speed.

Then you look at the speed currently indicated. If that speed is far less than 60 miles per hour, you increase the pressure of your foot on the accelerator pedal, perhaps flooring it if the difference is large. The car speeds up, and you keep watching intermittently until the speedometer needle begins to approach the 60 mark. As it gets within a few miles per hour of the desired speed reading, you begin to reduce your foot pressure, so the speed increases more slowly. Eventually your foot pressure comes to some steady level and the needle is stopped at the 60 mark. At that time, your car may be going 58.27 miles per hour according to your GPS device, but you are controlling what the speedometer tells you.

You are going at 60 perceived miles per hour.[3]

That is exactly how the cruise control in a car does the same thing, except that it uses a direct linkage instead of a foot, and a sensor connected to the car's drive shaft instead of a visual image of the speedometer.

When the car comes to a hill, the speedometer reading begins to drop, because the pressure on the accelerator pedal is not enough to sustain the same speed. You will probably press down just enough harder than before to keep the speed from dropping below 59 or 58 mph. If you're really fussy, you will end up pressing the pedal hard enough to keep the speedometer nailed at 60.0, as close as you can read it.[4]

When a tailwind springs up, the car will begin to go faster. You will reduce your foot pressure just enough to keep the car from going more than one or two miles per hour faster. You don't need to know whether a tailwind or a downslope caused the increase in speed; you need only to know that the speed is greater than the speed you want to perceive. If you're insistent about precision, you stop changing the foot pressure at 60.0 miles per hour. If a tire begins to go soft, creating a drag, and you're trying to calibrate your speedometer by timing the car as it passes mile markers, you won't quit changing your foot pressure until the speedometer says 60.0. A passenger in the rear seat opens a window, increasing the drag on the car. Speed: 60.0. Your tire reinflation system pumps the low tire back up to specs: 60.0. The roadbed gets rough: 60.0 (unless you decide to slow down).

The cruise control, a negative feedback control system, can do all these things, too, even though it can't sense upslopes, tailwinds, soft tires, open windows, or rough roadbeds.

3 If you were focusing on the speed indicated by the GPS device instead of the speedometer reading, the GPS would end up showing 60.0 miles per hour, and the speedometer would indicate 61.78 miles per hour. You are still going at 60 perceived miles per hour.

4 You will then be acting as an integrating control system.

This kind of controller can even be made to produce a trajectory of speeds. All that has to be done is for the driver to select a series of different reference speeds. The coast and accelerate controls of the cruise control let the driver change the reference speed. As each new reference speed is chosen, the controller will, all by itself, change the perceived speed upward or downward to match it. No equations have to be solved. Even a continuously-changing speed can be produced. If the reference speed changes too fast, the perceived speed will lag behind it, but that would happen with any kind of controller. A car can't change its speed instantly even if equations seem to say it can.

A negative feedback controller, or perceive-compare-act controller, or PCT controller doesn't have to know what is causing the speed to change. It doesn't need to know because its output action is applied directly to the perceived speed, and (within the limits of operation) can be varied enough to cancel the effects of any other influences, even novel ones, that tend to change the speed. Since this kind of system directly monitors the state of the controlled variable, and acts directly on that variable, it can skip all the complex computations required by the analyze-compute-act type of system. All you need to know is that pressing harder makes you speed up, and relaxing the pressure makes you slow down—as you perceive the speed.

If made from the same components and variables, a perceive-compare-act controller is far simpler than an analyze-compute-act controller, and as you will see in Chapter 3, a negative feedback controller is also faster than the MCT controller. The PCT controller can work even when unpredicted disturbances happen. It can work just as accurately as the MCT controller when the latter is constructed with real as opposed to imaginary components—and often more accurately, because MCT controllers can't compensate for unexpected changes in their own output actuators or in the plant. Most of what has been said in the literature of the life sciences about limitations of negative feedback control is simply false. In fact, the limitations do apply to the other kinds of control. How can you resist a disturbance of a kind that has never happened before? An MCT controller can't do that. A PCT controller can, because it senses and acts on the consequences of the disturbance, not on its causes.

For all I know, the MCT concept is a giant leap forward in control theory—for building artificial control systems. I'm convinced it is the wrong model to apply to living control systems. And I'm not too convinced it is a very elegant way to control anything. Thousands, I admit, may think otherwise. But history tells us there's always a chance that thousands can be wrong.

SIMULATION AND MODELING

In chapters to come you will be seeing many simulations of control processes. Some of these are interactive and show how your own control systems work. Others are general models showing how specific control systems might be organized, and how they would then work. These are all computational models in which behavior comes out as a sequence of numbers rather than real objects doing things (this would be true whatever approach to control theory is used). In some models the numbers are converted to pictures, like pictures of a moving arm. The pictures are there just to show where the numbers come from; the control systems being modeled, however, understand only the numbers, as neural networks in the brain understand only trains of neural impulses.

Note: The calculations done in this kind of model describe physical processes that the modeled system is carrying out. It is not proposed, however, that the modeled system carries them out by doing these calculations. It does them according to physical and chemical processes that work without mathematical calculations. We may model those processes mathematically, but the system we are analyzing just does them.

It may be helpful to have some idea of how such computational models work. A simple example can be seen in a mathematical model of a faucet from which water is flowing into a bucket.

To construct this model you have to represent some properties of the real system with equations that show how the physical variables are related to each other. A variable is just what it sounds like: something that can vary. Consider the rate at which the water flows. Say the flow rate is a variable with values ranging between zero (faucet closed) and some maximum number (faucet all the way open). In conformity to programming customs, you can represent the flow rate by a multicharacter symbol rather than a single letter as is normally done in algebra. You can, in fact, use the symbol *FlowRate*, which rather effectively reminds you of the meaning of this symbol. Using a single letter such as f would be much less memorable. How many words do you know that start with f? How many do you know (now) that start with FlowRate?

If you use multicharacter symbols, the convention of the implied multiplication sign between two symbols does not apply. "ABE" in algebra means A times B times E. The asterisk (*) is used to indicate multiplication in programming languages, so A times B times E looks like this: A*B*E, and ABE is just a U. S. President's name.

Now, define a second variable, called *HandleAngle*, measured in degrees. With a little experimentation using the real faucet and calibrated containers, you might find a handy relationship between flow rate and the faucet's handle angle:
FlowRate = 0.01 * HandleAngle. With the handle (the round kind) cranked open by two full turns, or 720 degrees, this equation would tell you that the flow rate is 7.2 liters per minute.

Suppose now that the bucket capacity is BucketCapacity = 10 (liters). If you turn the faucet handle to an angle of 360 degrees, how long will it take the bucket to fill?

The length of time, in minutes, needed to fill the bucket is BucketCapacity divided by FlowRate. FlowRate is 0.01*HandleAngle. Therefore you can calculate as follows (units of measurement indicated in parentheses).

HandleAngle = 360 (degrees)
BucketCapacity = 10 (liters)
FlowRate = 0.01*HandleAngle = 3.6 (liters/minute)
TimeToFill = BucketCapacity/FlowRate = 10/3.6 = 2.8 (minutes)

You don't even need a formal definition of TimeToFill. Using well-chosen multicharacter symbols makes these computations readable and suggests the correct definitions. I point out once again that the bucket and faucet do not do any of these computations. The computations are a systematic and quantitative description of the physical processes.

So far this is the approach taken in physics, of setting up a system of equations and then solving them mathematically to find the values of variables you're interested in. There's another way to solve these equations, however, which is to simulate the filling of the bucket by "acting out" the equations. This is much closer to the way the physical entities actually operate. Set HandleAngle to 360, and set WaterCollected = 0 and Time = 0. Then let water flow into the bucket for 0.1 minute. At the end of 0.1 minute, at a flow rate of FlowRate liters per minute, 0.1*FlowRate liters of water, or 0.36 liters, will be in the bucket. If you do this calculation over and over, keeping track of the elapsed time and stopping the calculation when there are 10 liters or more of water in the bucket, you can read the time and see how long it took. Here is a computer program to do all that (caution: the meaning of the equal sign is different now; to be discussed soon).

```
HandleAngle = 360 (degrees)
FlowRate = 0.01*HandleAngle
BucketCapacity = 10 (liters)
WaterCollected = 0 (liters)
Time = 0 (minutes)
repeat                                          (Start of program loop)
    Time = Time + 0.1
    WaterCollected = WaterCollected + FlowRate*0.1
until WaterCollected  >=  BucketCapacity        (End of program loop)
```

```
print Time
end
```

(The symbol >= means "greater than or equal to.")

The first time through this program loop (between repeat and until) the amount of water collected is 0.36 liters and the elapsed time is 0.1 minutes. The second time 0.72 liters of water have been collected and the time is 0.2 minutes. The 28th time around the loop, 28*0.36 or 10.08 liters have been collected (exiting the loop because that's more than 10 liters), and the time is 2.8 minutes. The printer will be stirred to print a sheet of paper with "2.8" on it, and the program will end. You could get more precision, of course, by making the time steps smaller than 0.1 minute. and carrying out the arithmetic to more decimal places.

Notice the peculiar programming statement, Time = Time + 0.1. If you interpret that as algebra, it is simply a false statement. No variable is equal to itself plus 0.1. In a programming language, however, the equal sign used that way doesn't mean equality: it means replacement. The value of the variable on the left is replaced by the value of the whole computation on the right. So Time = Time + 0.1 means that you take the previous value of Time, add 0.1 to it, and store the result back where the variable Time is kept in computer memory, *replacing* the previous value. Because this meaning is so very different from equality, programming languages use two different symbols. Pascal uses := to mean replacement, and = to mean actual equality. The C language uses = to mean replacement, and == (two equal signs) to mean equality. Almost every computer language has a way to make this distinction.

Since I'm mainly a Pascal programmer, I'll use := to mean replacement. So imagine adding that colon in all the preceding program steps and text where there is an equal sign by itself.

It would be easy to "solve" this equation to see how much water would be in the bucket after a given amount of elapsed time. All you have to do is pick some specific ending time ("EndTime"), change the until statement to read

```
until Time >= EndTime
```

and then print out the value of WaterCollected instead of Time. Simulations are much easier to use than analytical mathematics if you have a computer to run them.

You have now seen the difference between a mathematical simulation and an analytical solution of the system equations. To solve the set of equations analytically, you apply the rules of mathematics to obtain one of the variables as a function of all the others. Then, plugging in the values of the other variables, you can calculate the value of the unknown variable just by evaluating a single expression once. This is the approach behind the analyze-compute-act model. On the basis of an analysis of the physical situation, you compute the action needed to achieve some end-result, and then you carry out the action.

A mathematical simulation acts out, step by step, each relationship defined in the mathematical description of the system. You have to give some variables an initial value because each time around the loop that is part of every simulation, the next values of the variables are computed from the previous values together with whatever changes take place during the time represented by one iteration of the loop. The final result is obtained little by little just as the actual behavior being modeled changes through time from a starting condition to an ending, or continuing, condition.

The simulation looks much less efficient than the method of just solving the equations. It is—but it will be finished long before you have found a way to solve the equations if they are complex or you can't solve them at all. The simulation can run faster than real time, so you can see the effect of changing variables or constants as you change them, without having to pause to solve equations. You don't need to make any approximations to the observed relationships to make the simulation work, as you usually must do to find analytical solutions. Analytical solutions can get very lengthy and messy, and sometimes they simply don't exist. Then the simulation method is by far the quickest way, sometimes the only way, to get a right answer. And it has one more great advantage: it gives you an immediate picture of how the whole system behaves through time, because a simulation imitates the behavior of all the variables in the system at the same time.

This is how a simulation works, and I have a strong hunch that it is also how the brain generates behavior.

It's easy to experiment with simulations. Just for fun, let's say that there is a hole in the bottom of the bucket with an area of HoleArea square millimeters, and that the leak rate out of this hole is 1.0*WaterCollected*HoleArea (liters per minute). The higher the water level in the bucket, the faster it leaks out, and the larger the area of the hole, the faster the water leaks out. You can then change the statement FlowRate := 0.01*HandleAngle to read

FlowRate := 0.01*HandleAngle - WaterCollected*HoleArea.

You can change «FlowRate» to «NetFlowRate» if you like (it would have to be changed everywhere it appears). To plug the hole, just set HoleArea := 0.

Using a hole area of 0.4 square millimeters and a handle angle of 360 degrees, see what amount of water is in the bucket after an infinite time (you won't have to go on that long). You will notice a problem common to all models if you make the leakage rate less than the inflow rate by a large enough amount. Suppose the bucket can hold 10 liters. You can't get more than 10 liters into a 10 liter bucket, but the program thinks it can because we didn't tell it it can't. Modelers just have to think of graceful ways around such problems: for example,

if WaterCollected > 10 then WaterCollected := 10.

Finally, suppose the bucket is shaped like that bear-shaped container that honey is sold in. About the only way to represent the height of the waterline would be to experiment with the bucket and make a table of heights as a function of WaterCollected. It's really the height of the waterline that determines the water pressure where the hole is, and thus the leak rate. We've left the analytic method behind.

Now you know how all the simulations in this book work. They start with mathematical statements that describe how various parts of a control system work and how they affect other parts (as the flow from the faucet affects the amount of water collected in the bucket from moment to moment). Initial conditions are set by giving starting values to as many variables as necessary. Then the equations, interpreted as replacement statements (colon-equal), are evaluated one after another to compute the next values of all the variables. This process is repeated over and over, the system "behaving" until some condition is reached or until you stop it. In most of the simulations to follow, the time represented by one iteration of the program loop is 1/60 second. Modern home computers can do millions of computations in that length of time.

You don't actually have to know any of this about simulations to appreciate the chapters to follow. Most people, however, like to have a little meat on the bare bones of an idea, perhaps, as my old colleague the late Robert Kenley Clark liked to quote from Gilbert and Sullivan, "to lend an air of verisimilitude to an otherwise bald and unconvincing narrative." If you didn't find the bucket example interesting, you have my permission to forget it, if you can.

SOME PHILOSOPHY OF SCIENCE

As a final preparation for the simulations and experiments that follow, we need to look at some basic assumptions of science as they relate to the behavior of living systems. One of the principles adopted by biology and other life sciences has been that organisms are made of matter and therefore obey all the laws that govern the behavior of matter. While this is true, it is also untrue. It's not true that they obey ONLY the laws that govern the behavior of matter. It's true that a lump of solid matter will, because of gravity, fall to the ground when released, accelerating just as Newton said it would (but for air resistance). It's also true, however, that if the matter is shaped into a fuselage and wings, it will fly when you drop it. Laws that apply to lumps of material do not necessarily apply when the material is organized in ways other than lumps. If you analyze an airplane chemically or physically, it will prove to be made of materials that can't, in their raw form, fly. Yet they can be organized to fly.

Therefore we can conclude that the way material objects behave can't be predicted from their physical or chemical properties alone. There is something of nonmaterial nature called organization[5] that endows material objects with properties they would not otherwise have. Or to say that better in another way, the way that collections of materials behave depends as much on how they are organized as on what the materials are. Boats have been built from hollow logs, reeds, bamboo, canvas, iron, and concrete (but not unfrozen liquids or gases). You would never find "floats" on a list of the physical and chemical properties of these materials. Whether the material floats or flies depends on how you organize the materials, and in which liquids you try to make the materials float, or how dense the medium is in which you want them to fly. Penguins fly wonderfully, but only under water. Even a boat built of hollow reeds will eventually sink in water if you don't waterproof it, as Tor Heyerdahl discovered. Balsa-wood airplanes can easily be made to fly, though the airplane kit dumped out of its box will simply fall to the ground (same materials, wrong organization for flying).

The principle that organisms obey the laws of matter has therefore been more misleading than helpful. It has led to thinking that the same causal laws apply to living systems as to nonliving ones. While that is true at a certain level of observation—you can cause a mouse to fly by throwing it into the air—it is not true at any higher level. You can't cause a mouse to jump into the air *by itself* if it doesn't want to.

Under the traditional principle of lawful behavior of matter, wanting seems meaningless. In certain cases, that's the truth. It's not true that

5 Organization is expressed in terms of relationships among objects or parts of objects. Relationship is a nonmaterial aspect of things. The aspect referred to by between or above is not made of matter or energy.

a hot-air balloon rises because it wants to approach Heaven or that a cannonball falls because it wants to contact the Earth (as engineers used to think 500 years ago or so). Not all organizations of matter want things to happen or to exist. But other organizations do. Living systems do because control systems do, and living systems are made of matter organized as control systems.

So there you are: you are exploring, in this book, the properties of an organization of matter known as a negative feedback control system. Negative feedback control systems have, or can be given, intentions, desires, wishes, and preferences. They can act by material means on the material world to cause perceptions to appear and change until they match what is intended, desired, and so on. This does not require prediction (although prediction is sometimes useful for certain kinds of control). It does not require the future to affect the present (although a reference condition does specify a future condition that does not yet exist, just as a blueprint specifies the shape of a house that has not yet been built). It does not involve probabilities to any significant degree (given the blueprint, the house will almost certainly come into being almost exactly as specified, even if the owner runs out of money for a while or the builder finds he has run out of nails, and even if the architect scratches something out or adds something to the blueprint). The power and precision of control is evident everywhere around us; in any center of civilization almost the only things you can see or touch (unless you're outside and look straight up) are exactly what someone intended that they should be, including their shape, color, and function, not to mention the flavor and price of some of them. The world is packed so full of the consequences of human control behavior that they are invisible, utterly taken for granted, including not only the building in which the human-designed cages full of rats are studied, but the pencil and paper that the experimenter uses —exactly as he intends—to record the results of the experiments, and the journal in which the results are published.

Control, like digestion, is something everyone does, but hardly anyone understands. Let us see now if something can be done about that.

Chapter 2: Human Control Systems

We aren't going to spend any more time driving cars or generalizing. Instead we'll focus on capable and precise control systems that each reader carries around—a few among many hundreds, or thousands, that make up the human system. As you go through the rest of this book, you'll encounter updated examples of various PCT experiments and demonstrations that I've worked out since the early 1970s. I built the models to show how control systems work by actually running and behaving like control systems (and, demonstrably, like people). They run as they do because of the way they are organized, not because of labels or theories or technical jargon. The original models have all been rewritten and improved by Bruce Abbott.

The best way to learn about control systems is to experience them in operation, and the best ones to experience in operation are your own. You see your own control systems from a unique point of view unavailable to anyone else; you can see the real point of your actions where others observing you have no idea what you're perceiving. Only you can know (though others, and even a computer, can sometimes guess correctly) what you are really controlling. That is the point of the first demonstration.

ABOUT CONTROL and the MOUSE

It pays to use the practice sessions some demos call for until you feel comfortable and in pretty good control of the objects on the screen. Try big movements of the mouse first to see their effects on the display (not little tentative movements). Then focus on trying to achieve very accurate control, which takes a pretty lively kind of action to do.

- On the CD menu, select the first demonstration, ChooseControl, to see the screen pictured here:

Figure 2-1. Choosing what to control.

You see a roundish red object changing its shape, its horizontal position, and the direction of its axis. These changes are random but smoothed, the disturbances acting on each attribute following different time-patterns of change. All three of these aspects are affected at the same time when you move the vertical slider with the mouse (click and drag—left and right movements of the mouse are ignored).

- Experiment with the mouse to get an idea of how it affects the display. Also, watch what happens when you are doing nothing.

The task to perform in this step is simple. First, decide which one of the three variables (position, shape, or direction) you intend to hold constant. You could, for example, choose to keep the horizontal position at *centered*, or the shape at *spherical*, or the axial direction pointing *toward you*.

- Then move the mouse pointer to the slider, and start controlling.

The italicized terms are called reference conditions, presumably determined in your brain by the way certain reference signals are set (see below). Don't choose reference conditions too different from these; there are limits in the demos to the amount of movement that can be controlled; the display might get peculiar at extremes of change. The program will record the results and before the chapter ends, analyze them, but for now just experiment with maintaining the display in one or another of these three reference conditions.

- Try to control each of the three in turn. I'll wait here while you go and do that.

Controlling any one of these three aspects of the object is fairly easy after enough practice (shape is the hardest for me). You found that you could control only one at a time because all three were being disturbed, but by influences that varied independently. What you did with the mouse to counteract one disturbance did not counteract the others—it was just as likely to make the other aspects of the object vary even more.

Perhaps the most interesting thing about this demonstration is what you have to do in order to change from controlling one aspect to controlling a different one. You were looking at the same object all the time, so the change wasn't in what was visible on the screen. You were moving the slider vertically in an unpredictable (even by you) pattern all the time; the change was not in the kind of behavior you were producing. Nothing happened on the screen to say "Now it's time to control a different aspect of the object," much less tell you which different aspect to control. And it wasn't that you could see only one aspect at a time: all three were there and visible all of the time. The only thing that changed was which aspect you decided to pay attention to and hold constant.

You can look at different aspects of your environment and simply by deciding something, mentally, make the environment start behaving in some specific way in which it would not have behaved otherwise. Of course you can see how you are making it behave (or in this case, not behave), so there is no magic involved. But you can't see how you are changing what you pick out to control. You just do it. That's why we need models—they are intelligent (one hopes) guesses about mechanisms we can't see which would help explain what we can see.

AN EXPERIMENTAL RUN

Let's start to work toward trying out a model of this behavior. The first thing you have to do is get some quantitative data. Check the Status box, to the right of the Control Pad. The Status color is red and indicates practice. You have been practicing. Now there will be a real run during which the mouse movements and the disturbances will be recorded. When you click start (status changes to Started), you have three seconds to start controlling the aspect you chose to control; then the shape turns green as does the Status box, which reports: Recording. Don't worry if you're a little late; the results will not be affected much by a few extra seconds of not controlling. But try. If you're really late, click the ABORT button to terminate the run and erase the data so you can restart. Before you start, it will help a lot to be very clear in your mind about what you're going to control.

Hold the aspect you have chosen as constant as you can for one full minute (after one minute the data-taking terminates automatically and the recorded data is displayed; you don't have to watch the time). Go ahead. The result for a run by me is shown in Figure 2-2.

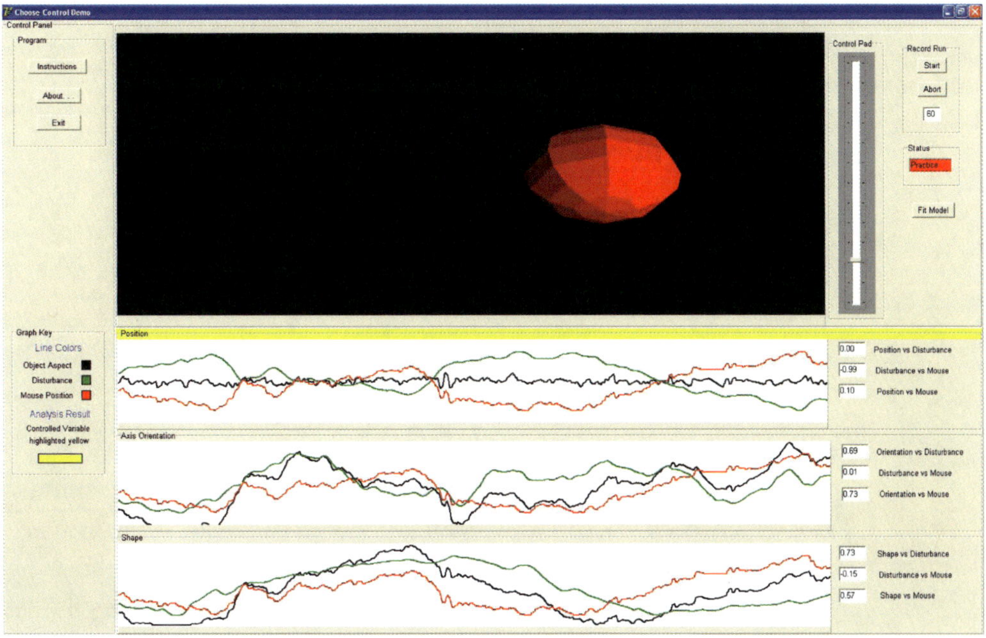

Figure 2-2. Screen after run. I was controlling the Position aspect, keeping the ball centered in its window. Note that the computer has detected which aspect of the ball was being controlled, and has highlighted the top of the correct rectangle in yellow.

Now each of the three rectangular areas below the ball, labeled Position, Axis Orientation (Direction), and Shape, shows three plots. Time advances from left to right, the duration of the plot being one minute.

The red trace representing mouse movements is exactly the same in all three plots. Up means the mouse moved away from you, or forward; down means it moved toward you.

The black trace shows how the Position, Axis Orientation, or Shape aspect of the ball changed during the run. The upward direction from center indicates that Position moved to the right, Axis Orientation tilted up, or Shape changed toward tall and thin.

The green trace in each plot shows the magnitude of the disturbance that affects one aspect of the ball. The effect of the disturbance and the effect of the mouse on each aspect add together. If the disturbance plot changes upward (positive) as much as the mouse plot changes downward (negative), there is no net effect on the controlled variable, the aspect you were controlling.

The first thing to check out is the constancy of the variable you were controlling. One of the black plots should vary the least, while the other black plots are varying by much larger amounts. If you didn't control consistently for most of the whole run, you'd better do it again; just mentally pick which aspect you're going to control, then click the START button and do the whole thing again. You need a consistent run to analyze, with no lapses of attention or sneezes in the middle of it, and no changes in what you were controlling.

In Figure 2-2, the black trace in the Position plot shows only moderate departures from a straight line. I was controlling Position. The other black traces show much larger variations. In your own demo, it should be easy for any onlooker to deduce which variable you had decided to control even if you didn't control it very well. In fact, even a computer can see what you were controlling; you will see that there is a yellow highlight at the top of the correct plot. If you go back and do the run again, deciding to control a different variable, the computer will show that the new variable was under control. *The computer is deducing which effects of your mouse movements were intentional.*

The method the computer uses to find the variable that is under control is to look for the maximum negative correlation between the mouse position and the magnitude of the disturbance over the entire run. In Figure 2-2, top plot, you can see that this correlation[6] is -0.99, which is an extremely high correlation (and with practice you can get it higher than that—perfect is -1.00). The large negative mouse-disturbance correlation in the Position plot shows that the mouse movements were accurately opposing the changing disturbance.

The corresponding correlations, much smaller, in the other two plots show that the mouse movements were not systematically opposing the disturbances, which had different patterns. (Note that this is also

6 All the correlations come from standard "Pearson's r" calculations, N = 3600.

a prediction of what you will see after your run, in terms of which correlations will be high or low. It's right, isn't it?).

A second way to find the controlled variable is to look for the minimum correlation between the disturbance and the aspect of the ball being influenced by the disturbance. For the controlled aspect in Figure 2-2, it is 0.0 (usually less than 0.1 or 0.2), while the corresponding correlations in the other plots are 0.69 and 0.73. This makes sense because the disturbance should have the least effect on the variable being controlled, and the most effect on the uncontrolled variables. A version of this approach can be used to identify controlled variables by applying disturbances to them, without even knowing how they are being controlled. If they don't change as much as they ought to, something else is opposing the disturbances, maybe a control system. This is known as "The test for the controlled variable" or TCV.

A third way is to look at the correlation between the mouse movements and the aspect being controlled. This is useful in cases where the disturbances are present but not known in advance or recorded. Here the correlation of mouse movements with the controlled aspect is 0.10; for the other two aspect plots, that correlation is 0.73 and 0.57. If you know how correlations are used in psychology and that does not surprise you, you are just reading these words without thinking about them at a higher level. What I just said was that the correlation between the mouse movements and the aspect of the ball that was being controlled is the *smallest* of the three corresponding correlations. Usually it's not as much smaller as in Figure 2-2, but it's always the smallest.

Yet the mouse movements are the only means you had of controlling the aspect you were holding constant.[7]

7 We have an action **A** that is affecting a controlled variable **CV**, which is also affected by a disturbance **D**. The correlation between **A** and **D** is very high (and negative); almost perfect. The correlations between **A** and **CV**, and **CV** and **D**, however, are very low: an apparent paradox.
<————————————- very high negative correlation ————————————>
Action <—very low correlation—> **CV** <—very low correlation—>**Disturbance**

Richard Marken has generated a striking illustration of this effect in his "mind-reading" demonstrations (www.mindreadings.com/demos.htm).

Along the way, this demonstration has solved a philosophical problem of some age: how to tell the difference between something a person does accidentally and something that is done on purpose. When the person is moving the mouse to control, for example, the Direction of the axis of the object, the mouse movements are affecting the Position and the Shape, too. If the task is to make the axial direction follow some predetermined pattern of change of the person's choosing (try it), it would be very hard for an onlooker to decide which of the change patterns visible on the screen was the result of a control process (that is, intentional) and which were just side-effects of that process. The standard way of determining dependencies would be to look for high correlations between mouse movements and the effects of mouse movements, but in fact the highest correlations in the long run would be between the mouse movements and the changes in the two *uncontrolled* aspects of the object.

So we have several ways of determining which of the effects of a person's actions are intentional and which are only accidental side-effects. We can do that in large part because now we have a way of saying precisely what we mean by *intentional*. Intentional behavior is control behavior.

Right now we have a record of a one-minute experiment in which you controlled one of three aspects of an object, including records of the disturbance pattern applied to each aspect and a record of the movements of your hand with the mouse under it. We now need a model that can behave just as you behaved during the run. By comparing the behavior of the model with your behavior, we can adjust the model until it behaves as much like you as possible. If the basic structure of the model is far from your structure, the best fit won't be very good. But if our guess about your internal organization is a good one, the best adjustments should make the model behave very nearly the same way you did. Please be patient now while the basic concept of a control system model is laid out. Then we'll try it out and see how well it can do with matching the data we now have (yes, if you didn't do it you're going to have to go back and actually do the experimental run if you want to follow what comes after the model is developed). We will now be looking at the basic organization behind perceptual control theory, PCT.

MODELING THE CONTROL SYSTEM

Figure 2-3 is a block diagram representing what appeared on the screen in this demonstration. In the diagram, the horizontal gray line is the boundary between you, the controlling system, (shown above the line) and its environment, the computer screen you were watching (shown below the line). We can see the environment; we know all about it, so we can model it easily. The small circle labeled INPUT QUANTITY stands for one of the controlled attributes of the object in the environment (the red object (ball) on the computer screen). Whichever attribute is being controlled is affected in two ways. There is a disturbance, which influences the state of the object. There is also an effect coming in from the right which can be traced backward through a box to the output of the system, a small circle labeled OUTPUT QUANTITY (the mouse position in this case). These two effects add to produce the state of the input quantity. The quantities vary smoothly rather than in jumps. The disturbance acts independently of the controlling system's action.

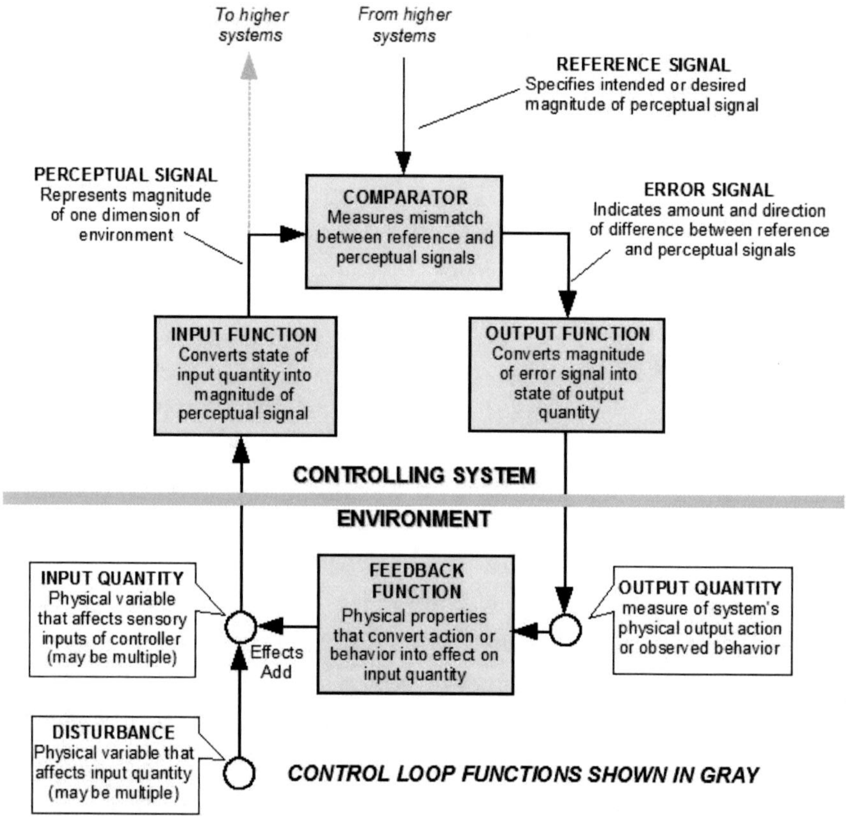

Figure 2-3. The basic negative feedback control unit of behavior on which perceptual control theory (PCT) is built. There are many of these systems at the same level, and many levels. The environment of higher-level systems includes lower-level systems. The effect of the output quantity adds to the effect of the disturbance on the input quantity.

The environmental part of the model below the line is as exact as we care to make it. Above the line, however, we must make hypotheses if this represents an organism. Above the line we have the controller, the system doing the controlling. No great ingenuity is required to decide that this system needs at least an input function (a way of sensing the state of the controlled variable) and an output function (a way of affecting it). It has to sense the variable under control to know whether its magnitude is less than or greater than the desired magnitude and of course the controller must affect the controlled aspect so it can be brought closer to the desired value (which requires a comparison). Let's look carefully at each component of the controller.

The Input Function

The sensing part at the input to the system is indicated above the line by a box labeled INPUT FUNCTION. Sensing is a very general concept, which includes not only sensing devices like retinal light sensors, auditory sound sensors, touch sensors, and balance sensors, but neural networks that can derive more abstract measures from multiple sensor readings. Just what we imagine to be inside this box depends on the kind of control process we're modeling. We can't really imagine much verifiable detail inside the box, because so little is known about the brain at this level of analysis. However, we do have one very broad hint about what the box accomplishes: it must provide the system with the same kind of perception of the variable outside the system that we, the human controllers, are experiencing and calling the controlled aspect of the object. In other words, the input function generates a variable neural signal that is experienced by the control system exactly as you and I experience whichever variable aspect of that object is under control: Position, Axis Direction, or Shape. We experience a lot that this system does not experience, of course, but we are modeling only what is necessary to explain the behavior in question.

We're not specifying which aspect of the red ball we are modeling here—it could be any one of the three. For each one, a different input function would be needed. Each input function would receive the same basic sensory information from the eyes, but different neural networks would come into play, deriving different kinds of information from the same stimuli. Since you can continue consciously experiencing all three of these aspects even though you're controlling only one, we have to assume that there are, in fact, three input functions involved here, each one extracting a different kind of information from the basic optical inputs, and all three operating continuously at the same time, in parallel. Just what determines the aspect that will be sensed and controlled at a given time is a question that looks ahead to a hierarchical control model, so we will skip that for now.

All of this discussion is going to boil down to a trivial result. The model we will use will say that the signal coming out of the input function and going upward, the perceptual signal, is equal in magnitude to our measure of the controlled variable just outside the system. The proportionality factor is 1: one unit of perceptual signal equals one unit of input quantity. All of the complexity in the input function is hidden in changing from physical units of measurement outside the system, below the line, to units of impulses per second inside the system. The perceptual signal consists of some number of impulses per second traveling upward in the line labeled perceptual signal. The magnitude of this signal (measured in Nervous System Units as impulses per second) co-varies with the magnitude of the input quantity measured in the appropriate physical or geometric units. That is what is meant by sensing the input quantity. The perceptual signal is like a meter reading. Note: meters display readings even when nobody pays attention to them.

The Output Function

On the output side, the output function shown in its own box represents the means this system has for causing changes in its environment. This output function, too, can be either simple or complex depending on the kind of behavior we're modeling. In the present case, it includes a hand holding a mouse (but not the mouse itself); muscles that operate the arm and hand, and some set of neural networks that route and apportion neural signals that cause the hand to move. While it would be possible to go into considerable detail about the output function (quite a lot is known about the properties of muscles and spinal reflex systems), for purposes of modeling this specific behavior we need only an overall picture of how a difference between the actual and desired states of the controlled variable gets turned into a change of position of the hand. We need a model that proposes a way for the value of an error signal (to be defined shortly) to be converted into hand position, the output quantity of this system. In the input function, a physical input quantity is converted into a neural perceptual signal. In the output function, a neural signal is converted into a physical output quantity. The varying output quantity is a measure of the action or behavior of the controller.

THE COMPARATOR

The final component above the line is the part of the controller where the actual value of the perceptual signal is compared with the intended value. The result of the comparison is a new signal called the error signal that represents the magnitude of the difference between actual and intended values of the perception.

The intended value of the perceptual signal is indicated by another signal of exactly the same kind as the perceptual signal, but with a different source. This other signal, the reference signal, is an example of the perceptual signal's state as it would be when the controlled variable is in its intended state. By itself it doesn't cause anything to happen, any more than a 30 MPH speed limit sign causes a car to go 30 miles or less per hour. It is simply there as a reference value, something against which the control system can compare its perceptual signal to see if it is too high (negative error signal, by convention), too low (positive error signal) or just right (zero error signal). The size of the error signal indicates the size of the difference; the sign indicates the direction.[8]

Even though the reference signal alone is not sufficient to explain any behavior, changing it will most definitely change the behavior of the system, just as raising the speed limit on signs usually results in cars going faster. The control system alters its own behavior according to the sign and size of the error signal, always in the direction tending to make the error signal smaller. By acting on the environment

8 Neural signals cannot change sign; there is no such thing as a negative number of impulses per second. In a real nervous system, control is always one-way: perception not less than, or else not greater than, the reference signal. What we model here as a bidirectional control system must really consist of two opposing one-way control systems, possibly with detectably different properties. Worth pursuing, when research demands that level of detail. For now, our models will be bidirectional and symmetrical.

below the line, it will make the perceptual signal change until it once again matches the reference signal. Because the reference signal is now different, the final value of the controlled variable outside the system will also be different—it will have just the magnitude needed to make the perceptual signal match (nearly) the new setting of the reference signal. Therefore changing the reference signal (with active cooperation from the components in the control loop) very reliably changes the magnitude or state of the input quantity, also called the controlled variable when we can prove it is under control.

THE FEEDBACK FUNCTION

We've reached the box in the environment labeled FEEDBACK FUNCTION. (It's time we saw the word feedback in this model.) The box represents the set of physical laws, properties, arrangements, linkages, by which the action of this system feeds back to affect its own input, the controlled variable. That's what feedback means: it's an effect of a system's output on its own input. You can't just give a control system feedback. It gives itself feedback. If you try to give it feedback from outside the closed loop, you will actually be applying a disturbance and the control system will resist its effects by pushing back, or if the disturbance makes the error less, by relaxing.

Notice that we classify the controlled variable as an input variable, not an output variable. The actual output of the control system is what its output function, or the actuators in it—the muscles—do to the immediate environment. In this case, the output function moves the hand, and since the hand is holding the mouse, we can measure the output by measuring the mouse position (or the slider position it affects). In this experiment, mouse position is determined almost completely by the behaving system; very little else interferes with it, so it's a good measure of the system's action. For output quantity in this experiment, you can substitute "mouse position in the Y direction." Y is the direction away from you on the table. Clearly you were not controlling the mouse to be in some preferred position, but *varying* its position as required to control the aspect of the object you were trying to hold constant. You positioned the mouse wherever you had to in order to counteract the momentary magnitude and direction of the disturbance.[9] Even you couldn't predict your own next output action.

We're now ready to analyze the data recorded during your one-minute test run.

PERFORMANCE OF THE MODEL

- Click the FIT MODEL button to call up the Model Fitting panel.

You are now seeing the behavior of a simulated control system organized as in Figure 2-3. A disturbance is being applied to the input quantity—in fact, exactly the same disturbance pattern that was applied to the aspect of the ball that you were controlling, as the computer identified it. The whole experimental run is being done in about 1/60 second, over and over, so you see a stationary pattern.

The red trace shows the mouse movements you produced during the experimental run. The green trace shows the model's simulated mouse movements. What we want, ideally, is for the plot of the model's mouse position to lie exactly on top of the plot of the real mouse position, your real behavior, for the entire minute of the run. Let's see how close we can get with this model.

There are two adjustable parameters in this simple model: the value of the reference signal, and a factor in the output function that determines how fast the output will change for one unit of error signal entering the output function.[10] For historical reasons, the second number is called the gain of the output function. It's also called the integration factor. You can use the up or down pointers in the two

9 In fact positioning the mouse requires another control system at a lower level, but the system we're discussing acts by varying, not controlling, the reference signal given to that lower system.

10 This is known technically as an integrating output function. The rate of change of the output depends on the magnitude of the error signal.

spinners to the right of the plot to run the values up or down, changing the value of the reference signal or the output integration factor.

As an aid to finding the best values, there is a box labeled RMS Error which shows the root-mean-square difference between the real and simulated mouse behaviors. Each of the 3600 differences is squared and added to a sum; the sum is divided by 3600 to get the average squared error, then the square root of the result is taken to get the RMS value over the whole run.

- Adjust the two parameters to get the smallest RMS difference, and look at the result.

This result is the main point of this chapter. Figure 2-4 shows how well the model fits the results of my run.

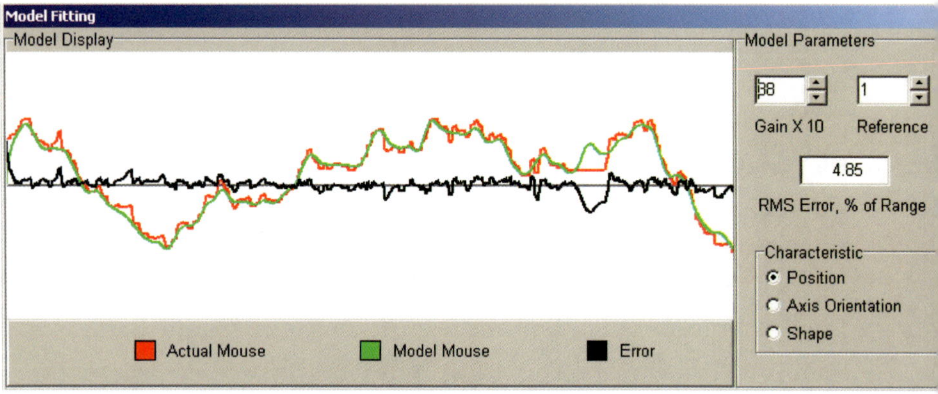

Figure 2-4. The new window shows the results of my run after adjusting parameters for best fit. The black trace shows the difference between simulated and actual mouse positions. The reference position, shown in the Reference box at the top right, is computed relative to the average value of the real mouse position. It's within one pixel of that value. The RMS error is 4.85 per cent of the peak-to-peak range of the real mouse position, a "10 - sigma" fit relative to the center value.

The final best-fit value of the reference signal shows how close to the specified condition you were actually aiming. The final value of the integration factor, the gain, tells us something about your own output function, a sort of global summary of its properties that gives a general view of how it works.

What you are seeing here is a working model, a simulation, being used as such things are commonly used in the hard sciences and engineering. You are now looking at the performance of the basic building block of Perceptual Control Theory, PCT. This unit, repeated to form side-by-side arrays of controllers and layers of control, becomes Hierarchical Perceptual Control Theory, HPCT. It is an idealized picture of a system which is no doubt different in detail, yet the actual forms, in the spinal cord at least, are quite like the organization of the basic control diagram. This model is no substitute for actual circuit tracing, but it gets us a long way toward being able to understand what the circuits that will some day be found accomplish.

The next chapter takes us a step toward a better model by introducing more parameters that can be varied, and is followed by a modern version of a control-system experiment that has been done thousands, perhaps tens of thousands, of times since World War II: the venerable tracking task. This will take us just about to the limits of what the few PCT researchers now working have achieved by way of matching models quantitatively to data.

ADDENDUM TO CHAPTER 2

We should not leave this chapter without discussing a different kind of negative feedback control model, proposed soon after the birth of engineering control theory. It is the model used by most engineering psychologists in the 1940s and 1950s, independently of Norbert Wiener's cybernetics movement. We will also look briefly at the place where a major cyberneticist's trolley went off the track.

THE HUMAN OPERATOR: STIMULUS-RESPONSE IN A CLOSED LOOP

In the 1940s and 1950s a movement called "Engineering Psychology" was launched, devoted to the study of human beings interacting with machines—machines like guns and airplanes, with military applications because of World War II. It is still around, having evolved into "ergonomics." I learned a great deal from engineering psychology about simulating control systems and about the mathematics of negative feedback control, and for a time considered entering that field instead of cybernetics, where there wasn't much engineering going on. But as I explored human control systems more deeply I came to the realization that engineering psychology had picked the wrong control model, even while analyzing control systems properly. I speak of the field as it was 50 or 60 years ago, but there are signs that these observations still apply.

This model contained most of the parts seen in Figure 2-3. Figure 2-5 shows a typical 1950s model of pursuit tracking in which a person uses a handle to make a cursor follow a moving target.

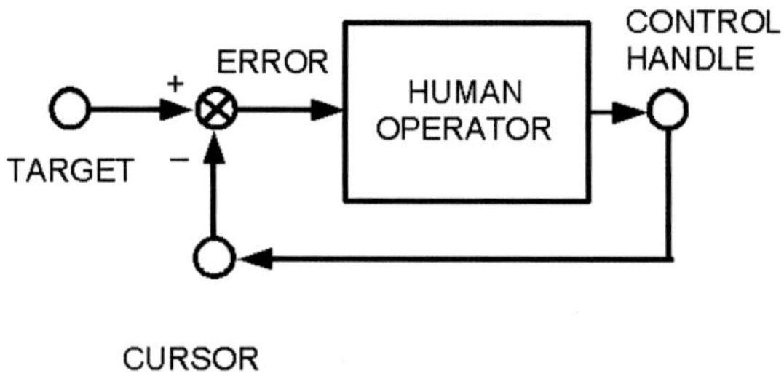

Figure 2-5. Engineering-psychology model of tracking behavior, ca. 1950.

The central idea in engineering psychology models was the HUMAN OPERATOR, described by a transfer function. This operator was embedded in an electromechanical system that formed a control system. The human operator itself, however, was not a control system, but an input-output or stimulus-response device, one component in a control system. This may have been a legacy from the mainstream psychology of the time.

The distance between TARGET and CURSOR is called the ERROR (the tracking error), which is an input to the HUMAN OPERATOR. The HUMAN OPERATOR transforms this input into an output in the form of the position of, or a force applied to, the CONTROL HANDLE, which moves the cursor. What makes this arrangement into a control system is the way the environment around the HUMAN OPERATOR is organized. The target is given the role that is played by the reference signal in Figure 2-3, and the comparator (the X-ed circle) is located outside the HUMAN OPERATOR, in the environment. This leaves the HUMAN OPERATOR occupying the position of the output function in Figure 2-3, converting error into action. The HUMAN OPERATOR is a stimulus-response or lineal input-output system, not a control system in its own right. The rest of the control system is in the environment.

A non-zero distance between target and cursor is an error only if the person doing the tracking wants the cursor to be on the target. In many tracking situations the tracker wants the cursor (in the form of a gun sight, perhaps) to lead the target by some distance, so the condition of "cursor on target" would then be an error, and the action of the control system must be such as to keep the cursor a fixed distance away from the target. How can this offset be put into the diagram? It has to go inside the HUMAN OPERATOR, to define the intended distance between the cursor and the target. And when that is done, we can see that the so-called error in Figure 2-4 is really just a perceived distance between two objects, and the perceived distance has to be compared against a desired distance inside the HUMAN OPERATOR. Then this diagram would work for any desired distance, including zero. With a few added details, the HUMAN OPERATOR transforms into the system in Figure 2-3.

Compute and execute: control as compensation.

A little after the engineering-psychology model was introduced, W. Ross Ashby, a prominent cyberneticist, described a negative feedback control system much like that in Figure 2-3, but ended up rejecting it as being inferior to another way of controlling which he mistakenly thought would work better. This compensatory model was claimed by Ashby to be superior to the negative feedback model, and apparently served to stimulate a number of other more detailed proposals that have evolved into what some now call modern control theory as discussed in Chapter 1. This approach looks simpler than the control system model, but that is only because most of the machinery needed to make it work is hidden.

When a disturbance is applied to something that is to be controlled, any kind of controlling system needs to apply an equal and opposite influence to the same something, if it's to be kept under control. Likewise, to create a specific change in the state of the thing being controlled, it's necessary to apply a series of precisely-adjusted influences in a specific pattern. The difference in architectures that we're talking about now lies in how the required actions are generated.

It won't hurt to say this again. In the negative feedback architecture, the action is created by detecting deviations of the controlled variable from its intended or desired state. Those deviations cause an error signal to appear, and the error signal is amplified or otherwise processed to produce the opposing action. Just how this happens and how it could work so quickly and precisely is not, as you will see in the next chapter, transparently intuitive, but clearly it does work that way, and very well. In all alternative architectures, the required actions must be computed from knowledge of the properties of the environment, exact observation of the disturbance and other variables, and precise calculation and execution of the right forces. All of that requires extensive knowledge and computation.

Ashby proposed that a control task can be done by sensing the state of the disturbing variable, computing how it would affect the controlled variable, and then computing and generating a second force coming out of the controller that would have an equal and opposite effect to oppose the effect of the disturbing variable, plus whatever other amount, direction, and timing of effect would be required to produce any new desired state of the variable.

Here is what he proposed:

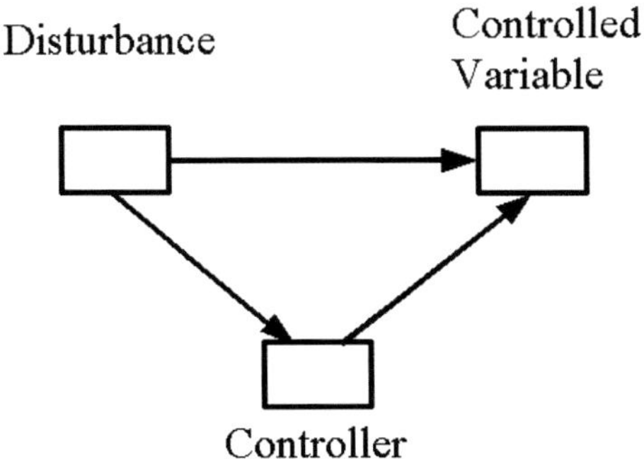

Figure 2-6. Ashby's preferred proposal for a control system. The controller produces an output (based on sensing the disturbance) that cancels the effect of the disturbance on the controlled variable.

This looks like a very simple scheme because it hides the computing equipment and whatever else is needed to acquire data, analyze it, and calculate and generate the output effects.[11] Modern control theory made some of those computations explicit using the Kalman Filter, invented in 1960. Of course this scheme would work to some degree if all the necessary resources and information paths not indicated were available. When the states of the variables are expressed as small whole numbers it would seem to work in imagination perfectly and instantly. If the disturbance produces 5 units of effect on the controlled variable, and the controller produces (with a properly-timed delay) -5 units of effect, it follows that the net effect on the controlled variable is 5 - 5 or 0 units to the same number of significant digits (one): the disturbance seems to be completely canceled. Ashby pointed out that even in a very good negative feedback control system, the disturbance always has a little effect, causes a little error, because the system's action is based on the error (using, of course, more than one significant digit to express the measurements). But his kind of system, he claimed, works instantly and without error. In terms of integer arithmetic, it does. But the real

11 One is reminded of a popular cartoon in which a series of equations on a blackboard includes a step called "... and then a miracle occurs."

system would not, as any engineer could have told him (Ashby was a psychiatrist).

If everything that is needed to make Ashby's model work were to be shown explicitly in the diagram, the resulting model would be extremely complex, far more complex than the negative feedback control system. Hundreds or even thousands of lines of program code would be needed as part of the model, and one would have to explain how and where in the brain all this computing is happening. The illusion that it can work perfectly is a result of ignoring all practical problems of building a physical device that can actually carry out all these hidden processes with infinite precision in zero time. In fact, starting with a given set of input and output components, a negative feedback control system will generate behavior that is faster, more precise, and more resistant to disturbances, while being simpler in construction that a compensatory system like Ashby's which uses the same components.

Here is *all* of the source code for the part of the program that runs the simulation for the model-fitting in the ChooseControl demonstration. The constant LastData has the value 3600. The variables adjusted to get the best fit are Ref and Gain. The constant dtime has the value of 1/60 (second).

```
Output := 0.0;
for I := 1 to LastData do   {repeats everything between begin and end}
begin
    perception := Output + Disturbance[I];
    error := Ref - perception;
    Output := Output + Gain*error*dtime;
    ModelMouse[I] := Output;
end;
```

Chapter 3: A Feel for Control

Now we have several immediate objectives. The first is for the reader to become better acquainted with the odd properties of PCT's basic control-system unit of organization, as we add some details to the model and then experiment with it in simulation. The second objective, addressed in Chapter 4, is to go back 60 years or so and replicate a venerable experiment. In Chapter 5 we shall correct a mistake that was made and prepare the way to broaden our picture of human control systems.

The important thing to accomplish in this chapter is to develop a feel for the way a control system operates, which may ward off some misconceptions that somehow got attached to this subject early in its development. Someone who had eavesdropped on conversations among engineers but obviously had never tried to build a working control system made some very confident claims: an input-output system that produces regular effects in response to stimuli (or to central pattern generators) is simpler than a negative feedback control system; it can behave faster, is more stable, is more precise, and is more likely to have evolved than the "complex" control system. Not one of those claims is correct. Once concepts like that have been proposed by someone famous and have been believed however, it is devilishly hard to get people to unbelieve them, even when given overwhelming evidence that the claims are mistaken.

I shall not waste your time trying to prove that such beliefs are mistaken; I shall simply show you that they are. Whether that is sufficient is beyond my control.[12]

A control system is an organization of parts. Each box in a control system diagram represents one physical part of the system as a cause-effect device. The input function discussed in Chapter 2, for example, might represent an eye, which receives images as input and emits neural signals as output representing the amount of light falling on rods or cones of the retina. As the inputs vary, the output signals vary. In contrast, the boxes in a flow chart (for example, a diagram made of labeled boxes and arrows showing how to assemble a toy) show activities being carried out one after another; what goes on in each box may be carried out by the same physical system represented by other boxes, but doing different actions. The TOTE[13] unit is a flow chart, for those familiar with that famous idea: only a flow chart could have a box labeled Exit. One does not exit from a control-system diagram; it is always present. And an input function can never serve as an output function: an eye is not a muscle. The parts of a system diagram are physically separate. Even if you're not looking at the input function as you examine the diagram, the real input function is always there and working. A system diagram of the kind we use here represents a physical organization; a flow chart represents a series of behaviors. Understanding this difference may save you some confusion.

The intuition trained into all of us says that first the stimulus occurs, then the response. But in a control system (as in real organisms), the environment is always stimulating sensory nerves, and motor actions are always affecting sensory nerves through external feedback connections, so it is almost impossible to apply an external stimulus that is not greatly modified, even canceled, before it gets to the sensory inputs. We actually saw this principle in action as the block diagram in Figure 2-3 of Chapter 2, but I said nothing about it then.

12 "A man convinced against his will/ is of the same opinion still" (Wall hanging embroidered by my grandmother Alice before I was born).

13 Test-Operate-Test-Exit

Now we are going to look at an expanded form of Figure 2-3—not greatly expanded, but expanded enough to improve the predictions the model can make, and also embellished enough to let more experiments be done.

AN EXPANDED MODEL

If you haven't already done so, this would be a good time to exit the previous program (use the EXIT button) and pick the next item from the menu, Demo 3-1.

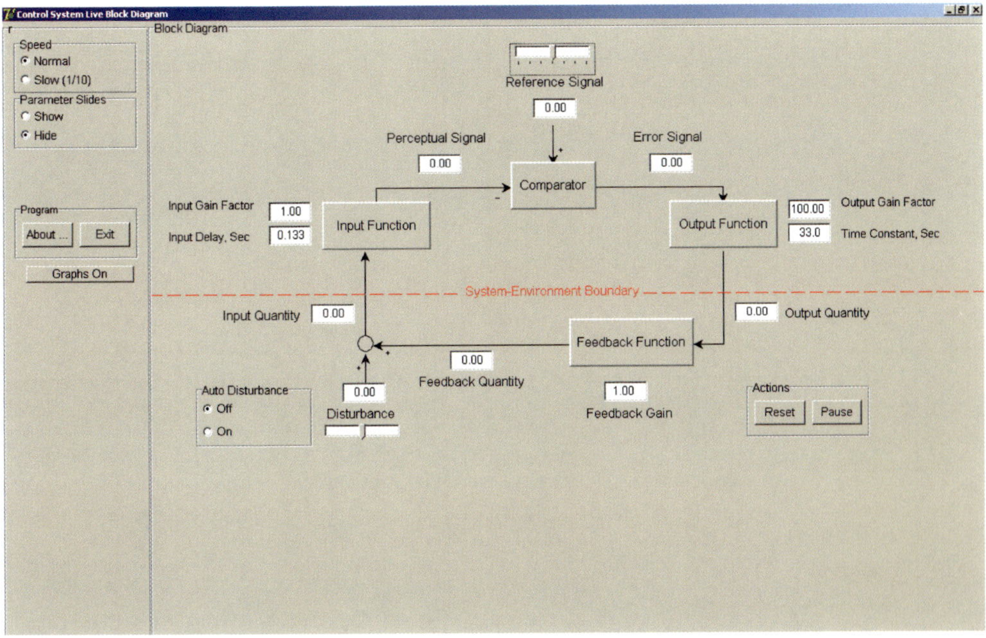

Figure 3-1. Live Block Diagram of a control system with adjustable disturbance and reference signal.

On the screen you see the block diagram shown in Figure 3-1. There is one slider for the disturbance and another for the reference signal. These are the two main influences on the control system from outside the control loop. There's a button for turning on a graphical plot, but leave it alone for now—it would just be a distraction. Numbers next to each variable indicate the magnitudes of the disturbance, input quantity, perceptual signal, reference signal, error signal, output quantity, and feedback quantity. They are all zero at the moment.

Behind the diagram is a working model of a control system, a simulation program that is recalculating all the values of the variables in this system sixty times per second. The perceptual signal is computed from the input quantity times the input gain factor (1.00), delayed by the amount of input delay (0.133 second). The error signal is computed from the reference signal minus the perceptual signal. The output quantity is computed from the error signal amplified by (multiplied by) 100 and slowed by a time constant of 33 seconds. The input quantity (where we started) is computed from the output quantity times the feedback function, plus the disturbance magnitude. I repeat: as you watch, all these computations are being done sixty times per second, producing the numbers for signals and quantities.

There is zero amount of disturbance, and the reference signal is set to zero; as a result, all the other variables are zero, too. So is there any control system? Is a system that is not doing anything controlling?

The answer lies in the organization, and it is Yes. This is a live, active control system, but you can't tell that until you disturb it. If it is not controlling, it will do nothing when you disturb it. So, disturb it.

- Click-drag the disturbance slider and bring the magnitude of the disturbance all the way to the maximum positive value of 15 (slider all the way to the right).

As you change the disturbance, you will notice the input and output quantity, as well as all the signal values, changing. If nothing had happened when you changed the disturbance, this system would not be controlling. But something does happen. Look at the number beside the label OUTPUT QUANTITY. This is the measure of the system's physical action on the environment, its behavior. As soon as you start to change the disturbance to affect the input quantity, long before you have finished changing it, the output quantity starts to change. Because of that, the feedback quantity effect on the input quantity begins to change, so the net effect on the input is not the effect of the disturbance alone. It is the combined effect of the changing disturbance and the changing output, the action of the control system.

When you change the disturbance to +15, the output quantity changes, soon, to -14.85. When you change the disturbance to -15, the output quantity changes to +14.85. In either case there is a final difference of plus or minus 0.15, as you can see from the number beside the INPUT QUANTITY label. Since the input quantity's magnitude is the sum of the disturbance magnitude and the magnitude of the output quantity, that is what you would expect. The input quantity remains near zero because the output quantity changes oppositely to the disturbance, and very nearly equally. You may think you're disturbing the input quantity of this control system, but except for a short time after the disturbance begins, you're not. The control system is keeping you from stimulating it—from disturbing what it is sensing. Without the feedback from the output to the input, a disturbance of 15 units would cause the input quantity to have a magnitude of 15 units. With feedback, the magnitude of the input quantity is only 0.15 units, after the changes have settled down. Ninety-nine percent of the effect of the disturbance has been canceled.

Suppose now that you are looking at the real system represented by this diagram, but do not know of the model above the line. Suppose also that you do not know of the input quantity itself, or if you do measure it, you dismiss it because it changes so little. Suppose you know only what you did to the environment near the system (the change in the disturbance) and the action the system produced immediately afterward (the change in the output quantity). If this were an organism instead of only a live block diagram, what would you conclude about the way this organism works?

A great majority of behavioral scientists concluded a century or more ago that organisms that behave like this are constructed to react to stimuli by producing responses. They guessed that a "distal stimulus" (disturbance) acts on the sensory nerves (the input function) which send their signals by various paths through the brain (input, comparator, and output functions), to the muscles which generate the movements that are called behavior (output quantity). So clearly, you could produce any behavior of an organism that you wanted if you could just discover which stimuli cause that behavior. That was the conclusion reached by John Watson, the father of behaviorism, 100 years ago. That was also a huge mistake, but I suppose that we will look at least that naïve to scientists reading this 100 years from now.

The theory that stimuli cause responses is not unreasonable. It is what any intelligent person might propose in the absence of any concept of a control system and its basic parts and organization. The theory of stimulus and response was created at least a century before control theory was invented (Watson got his idea from biologists), and probably a very long time before that, going back at least to René Descartes. Nobody could have known that the appearances which led to this theory are illusions.

Well, our model can be made to produce any amount of output you please (within reason) just by setting the disturbance to some specific value. So why isn't this a stimulus-response system? Seen from the outside, it certainly looks enough like one to have fooled several generations of scientists.

In the model, you can quickly see what is wrong with thinking of it as a stimulus-response system. Leaving the disturbance magnitude wherever it happens to be, look at the value of the input quantity. It is close to zero. The disturbance is hardly affecting the input to this system.

- Now go up to the slider at the top and raise the reference signal to its maximum positive value, +15 units.

You will see that the input quantity is quickly brought to nearly the same value. Since the disturbance (stimulus) didn't change, all that could have altered the input quantity was a change in the action of the system itself. The system produced a large amount of response with no change in the stimulus from the outside world at all. Look at the numbers.

Why don't we think of the change in the reference signal as a stimulus? The main reason is that in the larger model of which this diagram is a part, the reference signal does not arrive via a sensory receptor from the environment; it is generated by higher systems inside the organism. We, of course, are changing that reference signal from outside the control system. But we are playing the part of systems higher in the brain than this control system, systems which act by altering reference signals for lower systems. We are doing something that cannot be done from outside to a living organism (without surgery or chemicals)—we are reaching inside the organism to alter its reference signals directly.

I'll repeat this just to make sure the point gets across.

- Set the reference signal back to zero, and the disturbance to zero. Now quickly raise the disturbance, which acts directly on the input quantity, to +15, and notice how much the input quantity changes.

  Tip: Use the arrow keys for fine adjustments.

We're talking about the effect of a so-called distal stimulus on the sensory inputs of this system, via their effects on the input quantity. If you move the slider as fast as you can, there is a large initial effect on the input quantity, but very quickly the action of the system brings that effect nearly back to zero. If you change the disturbance more slowly, its effects will be much smaller.

A constant disturbance has almost no effect even if it's quite large.

- Now set the disturbance back to zero, and raise the reference signal to +15 units.

The input quantity rises immediately until it is very nearly +15 units and stays there. The value of the input quantity is clearly determined by the reference signal, which acts from inside the whole system, but not by the disturbance that acts from outside on the same input quantity.

Make sure you understand this, because it's the basic principle of perceptual control theory. The actual input to this system is the input quantity, the part of the physical environment directly sensed by the system. Disturbances of the input quantity, however produced, have only a transient effect which is soon canceled by the output action that immediately ensues. Slowly changing disturbances never do produce much effect on the input quantity. However, the reference signal has a large effect on that same input quantity: the result of raising the reference signal to any value is to bring the input quantity to nearly that same value. And if you now vary the disturbance again, you'll see that the input quantity stays in this new state, being almost undisturbed. The changing actions of the control system see to that. This system can not only cause its own input to come to a specified state, it can keep it close to that state despite arbitrary disturbances from outside the control loop. It does so, that is, if you're looking at the right input—the input that the system is controlling. The disturbance could be called an input, but it is not being controlled. Only the variable designated as the input quantity is being controlled.

We must conclude that the operation of this system gives the reference signal virtually complete control over the input quantity (and the perception of it), at the same time preventing disturbances from altering the input quantity in more than a transient way, as long as the system is operating normally.

You can't just "stimulate" this system by disturbing its sensory input. As soon as you try to do that, it alters its own effect on its own input and modifies the effect of your stimulus—drastically. This is why, in conventional behavioral experiments, you observe brief shocks and flashes of light and puffs of air and sudden loud sounds and relays clicking and in general "events" happening very suddenly. Unless you act very quickly, quicker than an organism can start to act, the organism will have just as much influence on its own sensory inputs as you have, and often a lot more influence. To do conventional experiments with behavior, you have to thwart the normal ability of living control systems to control what happens to them. In fact, you have to arrange matters so the living control system is forced to behave like a different kind of system, a stimulus-response system. You have to make the organism fit the theory instead of the other way around.

Why is there this asymmetry in the effects of the reference signal and the disturbance? They both enter the control loop from outside; just speaking geometrically, they seem to have about the same relationship to the control loop. It has been said (another rumor) that in this feedback arrangement, the organism affects the environment and the environment affects the organism, so you could say that either one "controls" the other. That's what happens when you look at a quantitative relationship qualitatively. It may be true that the controller's input function is the environment's output function and vice versa. But in fact this arrangement is highly asymmetrical.

The asymmetry is found in the gain factors of the input and output functions. Next to the output box are two labels: one says that the output gain factor is 100, and the other says the output time constant is 33 seconds. The time constant is there to keep the system stable (later you can reduce it to see what happens), but the important factor in the asymmetry is the gain factor of 100. That number says that whatever the value of the error signal, the output quantity (after it has reached its final magnitude) will have a magnitude 100 times that large. Over on the input side, we see that the input quantity is converted to a perceptual signal with a gain factor of merely 1. In the 100:1 ratio of these two factors we find the asymmetry that allows the controller to control the environment without the environment controlling the controller.

In fact we just demonstrated that conclusion, by examining the effects of the disturbance and the reference signal on the input quantity. The value of the reference signal quite accurately controls the value of the input quantity; the value of the disturbance does not "control" the value of the controller's output quantity, which is the environment's input quantity. The disturbance affects the output quantity, to be sure—but the reference signal affects it just as much. The output quantity is not protected against disturbances.

Another term for input quantity, once you have shown that it is under control, is controlled quantity or controlled variable.

There is more that can be said about this simple model, but it gets more interesting if we open up the model to more adjustments. We can make this into a five-parameter model, and do some experiments which ought to surprise you.

- Click the **show** button in the Parameter sliders panel.

MODEL WITH ADJUSTABLE PARAMETERS

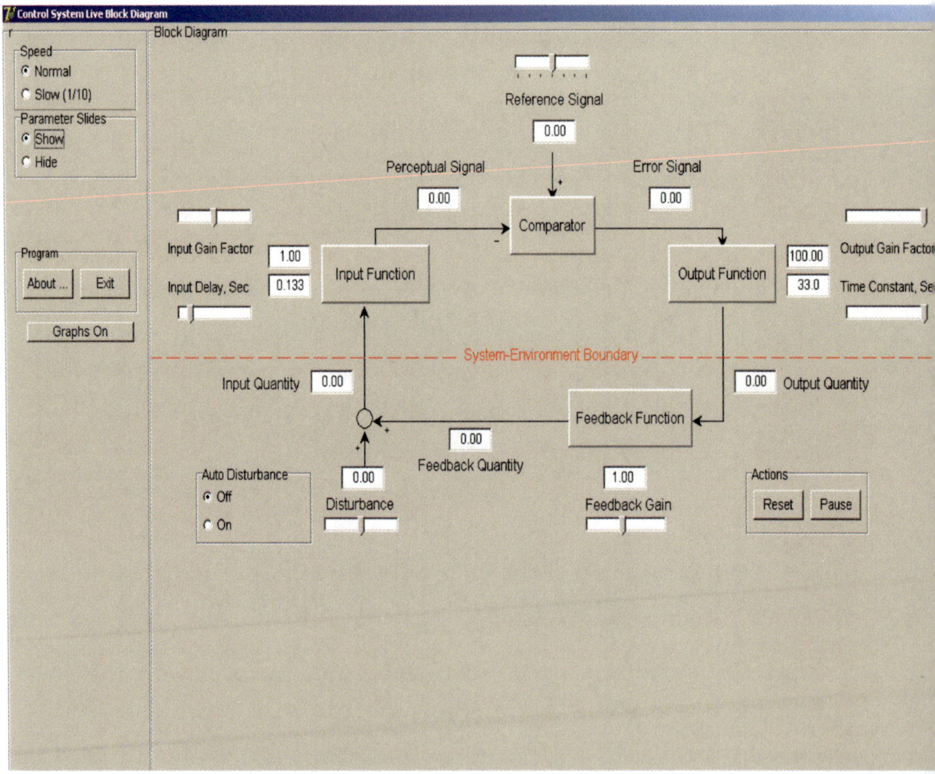

Figure 3-2. Live Block Diagram with parameter-adjustment sliders added. The lower part of the screen is the graph plotting area (graphs are off now).

Now five more sliders have appeared on your screen, as shown in Figure 3-2. We will discuss each one.

Input Delay. In all real systems there are time delays caused by the fact that signals take time to travel and computations take time to carry out.

This kind of delay is called a *transport lag*. It's similar to the delay of sound when you're watching and listening to a band a few hundred feet away. You see the percussionist's hands swing the cymbals silently against each other and then after a perceptible delay you hear the crash. When the crash is heard the sound is normal—the onset is just as fast as it would be without any delay. Nothing at all is heard of the cymbal until the end of the delay, and then a normal cymbal sound is heard.[14]

The Input Delay slider adjusts a transport lag in the input function—there could be another transport lag in the output function, but this one is sufficient to show the effects of transport lags in the loop. The least delay possible other than zero is 1/60 of a second, the duration of one iteration of the program (16.7 milliseconds). The 0.133 second input delay is about the amount of transport lag we find in real human visual-motor tasks, as you will be able to verify later. You may have heard that delays in control systems render them ineffective and unstable, so such models can't account for real motor behavior. You can judge for yourself whether this idea is true. Delays have to be compensated for in a proper design, but the way of doing so is quite simple (in this model, the integral lag in the output function does the required compensation).

Input Gain Factor. Just outside (below) the input function is some aspect of the environment that the system is controlling. It is measured in physical units. Inside the controlling brain, a perceptual signal is continuously being generated by the input function. It is measured in impulses per second, as PCT theorists currently conceive of signals in the brain. The input gain factor determines how many impulses per second of perceptual signal are generated for

14 There is another kind of lag, an integral lag, caused when changes are slowed. There is some effect of the event right away, with no delay, but with a slow build-up that makes it seem that there is a delay. The output function we assume in the live block diagram contains an integral lag.

one unit of the input quantity. It is currently set to 1, which actually determines what signal units will be used inside the system.[15]

There is a slider for adjusting the input gain factor to values other than 1. The chief reason for doing so would be to show how changes in the input sensitivity will affect the performance of the controller, as we will do before long.

Output gain factor. This is the conversion factor from Nervous System Units, NSU, to physical units of output quantity. For any one control system, the meaning of 1 NSU is set when we define the input function for that system.

The output gain factor determines how many units of output quantity will be generated by one unit of error signal entering the output function. It is set right now to 100. With this high a value of output gain factor, and with all the other factors in the loop being equal to 1 as they are, this would be a very precise control system, possibly more precise than any organism could be without artificial aid (though control systems using vision for feedback can be very precise). As we saw, even with a maximum disturbance of 15 units the biggest continuing error we ever saw due to the disturbance alone was 0.15 units, producing a precision of control of about 1 part in 100 in relation to the reference signal's magnitude. It's not a coincidence that with a gain of 100, the precision would be close to 1 part in 100. The rule is that with gain factor G, the precision of control is

15 Suppose the input quantity is the angle of the elbow joint. Just guessing, we might say that the neural signal representing joint angle is zero with the elbow straight out, and 300 impulses per second with the elbow bent to an external angle of 135 degrees (45 degrees past a right angle). We therefore have 2.19 impulses per second of perceptual signal for each 1 degree of external elbow angle. If we define one nervous system unit (NSU) of signal as 2.19 impulses per second, then the conversion factor from degrees of elbow angle to NSU becomes 1. This determines the scaling factor between physical units and NSU for this whole control system. Thus it's always possible to make the input gain factor equal to 1.

one part in (G+1)[16]. The main purpose of using such a high gain factor in the demonstration is to give the system nearly perfect control. This is the best way to show the basic properties of control systems, so you will recognize them later when we see less perfect living control systems in operation.

Output Time Constant. This is also referred to as a "slowing factor." Suppose we could isolate this function's output effects from the rest of the loop (the "open-loop" condition), and suddenly apply one unit of error signal to it while measuring its output. We would see an initial rise of output that slowed and slowed until the output approached a final limit very, very slowly. The time taken to reach 63% of the final value is the time constant. This time constant is now set to 33 seconds. If the loop were not closed, you would have to wait a long time just to see that the rise was not going to go on forever. You can see this happen if you first set reference and disturbance equal to zero, then set the feedback factor to zero and change the reference signal a little bit.

But you do not, in fact, have to wait that long when the loop is closed. The closed-loop time constant is about 0.3 second. This system is behaving about as fast as a real human being would behave. The observed time constant is the open-loop time constant (determined by a physical property of the output function that we can specify) divided by the loop gain. Notice that this control system can produce an action appropriate to the size of a disturbance 100 times as rapidly with the feedback loop intact as it could if it simply responded to the disturbance in an open loop with no feedback (and with an appropriate adjustment of output or input gain factor to produce the right amount of output). Feedback does not make control slower than simply responding to a stimulus; it makes it far faster. H. S. Black of

16 What actually matters is the product of all gain factors in the loop (including the one in the box outside the controller, the feedback function). The G in the above rule is the loop gain. Since all the gain factors here are 1 except the output gain factor, the loop gain is 100 right now. The factor for the effect of the perceptual signal on the error signal is -1 and all other factors are positive, so the overall feedback is negative.

Bell Laboratories discovered this fact in 1927 and put it to use in the design of transcontinental cable telephone repeater amplifiers, but it never got into the behavioral sciences.[17]

In the next chapter, we will find out what the actual transport lag, output gain factor, and output time constant are for a typical human being (you).

Environmental Feedback Factor: Below the Feedback Function that links the output quantity to the input quantity there is shown an environmental feedback factor of 1. This says that with the disturbance held constant a change in the output quantity of 1 unit would produce a change in the input quantity of 1 unit. If the feedback factor were 2, a 1-unit change in the output quantity would produce a 2-unit change in the input quantity. We are by no means limited to such simple kinds of feedback functions, but there is so much to talk about here that a simple linear function is enough to handle for now. Later we will see what happens when we put a mass on a spring with damping where this simple function is.

Effects of changing the parameters.

Control systems have more unexpected properties beyond those we have seen so far, among which is an apparent reversal of the effects of changing system parameters.

- Set up the system by adjusting the disturbance to zero and the reference signal to 10.0 units.

Reminder: the right and left arrow keys are useful for fine adjustments once you've clicked on a slider. Now we can consider the effects of changing some of the parameters.

17 Black, H.S. (1934, January). Stabilized feed-back amplifiers. Electrical Engineering, pp.114-120.

Changing the output gain factor.

As you can see by experimenting, no matter where you set the disturbance and the reference signal, the output quantity's steady-state magnitude is always 100 times the magnitude of the error signal, including sign (actually that's true to about 18 decimal digits, but the displayed numbers are rounded off to two places). So what do you suppose will happen if we cut that output sensitivity in half? Such a loss of response to a driving signal could happen in a muscle that becomes fatigued. Instead of 1 unit of error signal producing 100 units of output quantity, 1 unit of error signal will produce only 50 units of output quantity.

Try to reason it out before you actually move the slider to make the change. Right now, with the reference signal set to 10.00, the amount of output quantity is just enough to affect the input quantity sufficiently to bring the perceptual signal to 9.90 units; this makes the error signal's magnitude 0.10 units. Consider this specific question: If you now reduce the output gain factor to exactly 50, what do you predict will be the resulting magnitude of the output quantity after the effects of the change have settled down?

Here's one way you might go. If you reason sequentially, as seems reasonable at first, you will figure that cutting the output gain factor in half will cut the output quantity in half. If the output quantity is cut in half, the same would happen to the input quantity and the perceptual signal. The error signal, compared to its former value, would be enormous. So your first guess might be that control would suddenly become very inaccurate, hardly controlling at all.

But wait. If the error signal becomes enormous, that would mean we have an enormous signal entering the output function, which would be multiplied by 50, giving us a monstrous amount of output. Let's see, half of the previous value of output quantity (about 10) is about 5 units, so the input quantity and perceptual signal would be 5 units. With a reference signal of 10 units, that would produce an error signal, reference minus perception, of about 5 units. Multiply that by 50 and you get 250 units of output instead of the 9.90 units we had before. So reducing the output gain factor by a factor of two actually increased the output by a factor of about 25. Wow, what a surprise!

But wait. If the output goes to 250 units, then so does the input quantity and the perceptual signal. Now we have a perceptual signal of 250 units and a reference signal of 10 units, producing an error signal, reference minus perception, of—wow—negative 240 units. And that will produce an output of -240 times 50 or -12,000 units, twelve thousand units in the negative direction.

But *wait*—this is not working out. It looks as if reducing the output gain by a factor of two turns this control system into a runaway disaster, bouncing back and forth between extreme limits that just get more extreme every time around. It's time to see what really happens.

- Use the slider to reduce the output gain factor to exactly **50**, doing the fine adjustments with the right and left arrow keys on the keyboard.

What has happened to the output quantity? It has changed from 9.90 to 9.80, a drop of 1 per cent. It did not drop to half its previous value. It hardly changed at all. Despite losing half its sensitivity, this "muscle" is pulling almost exactly as hard as it was before.

But *what* has changed? Something must have changed, and more than just a little. That little drop in the output quantity reduced the input quantity and the perceptual signal by the same amount, so the perceptual signal is now 9.80 instead of 9.90. Hardly any change there, either. But, ah-ha, that little drop in the perceptual signal actually *doubled* the error signal! The error signal was 0.10 units, and now it's 0.20 units. There's the big change. It may be a small change numerically, but it's a very large change as a proportion of the previous value. And that change gets multiplied by 50 in the output function, so its effect is not trivial. When we halved the output gain constant, we can now see, the main effect was just about to double the error signal entering the output function, and thereby to prevent the output quantity from changing more than a very small amount, one per cent.[18]

Reasoning backward in circles.

Obviously, tracing out the effect of the change in the usual way, following cause and effect sequentially around the loop, did not work. How can we reason intuitively about this sort of system without actually solving equations?

The best way is to start by considering the whole system as one simultaneous unit. What is this system doing? It is keeping its perceptual signal, as a final condition, very nearly at the same magnitude as the reference signal it is receiving. So assume that final condition as the starting point.

- Raise the output gain to 100 so we start with the same conditions as before.

Suppose the final condition is "perception equals reference." Since the reference signal is 10 units, that means, to a first approximation, that the control system is keeping its perceptual signal at 10 units. Looking at the diagram, we see that the input gain factor is 1, so the input quantity must be 10 units, and because the environmental feedback factor is 1 and the disturbance is zero, the output quantity must also be 10 units. The output gain factor (if you remembered to reset it) is 100, therefore the error signal must be 10/100 or 1/10 (0.1)

18 Consider what a "simple" input-output system would have to do in order to show this same stability in the presence of changes in the system's actuator. First, it would have to sense or otherwise know about changes in the gain factor in the output function. Second, it would have to have a built-in computer programmed to compute inverses of various functions to determine how much the signal entering that output function, with its slowing factor, would have to be raised in order to produce the same output effect as before. Those sensing and computing functions would exist in addition to the input and output functions that we already have. Of course we assume the input-output system would otherwise have the same physical properties as the control system (including the delay), not counting the comparator which performs a simple subtraction. Conclusion: Control systems are far simpler than straight-through systems that do the same thing.

unit. It is actually 0.099—0.10 to two decimal places.

For the error signal to be 0.10 unit, the perceptual signal must be 0.10 unit smaller than the reference signal. The perceptual signal will therefore be 9.90 units instead of the 10.00 we initially assumed. We now have the right answer to two decimal places after the first complete trip backward around the loop. We can get even closer to an exact answer by going around the loop again (get out your calculator and try it). When the loop gain is lower, we may have to go around backward several times to get a sufficiently accurate answer.

The convergence to the right answer is very fast when you start with the assumed end condition and reason backward around the loop.

Going forward around the loop, following cause and effect naively, gives a very wrong answer.

It's necessary to reason backward in a circle around a control loop to get right intuitive predictions of its behavior.

Now you can try this for yourself by changing still another parameter in the loop.

Changing the input gain factor.

Be sure the output gain factor is exactly 100. The error signal's value will be 0.099, and so forth. The disturbance should still be zero and the reference signal should still be 10 units.[19]

Don't do it yet, but what you are going to do is increase the input gain factor from 1 to 2. Before you do it, try to reason out what the values of all the variables will be after the change: perceptual signal, error signal, output quantity, and input quantity. You should be able to guess the numbers exactly, since there are only a few significant digits shown. Start with the condition that this control system maintains, and then reason backward from it.

19 Notice how natural it is to specify what you should be perceiving instead of how you should act.

When you have your predictions, read on.

- To increase the input gain factor from 1.00 to 2.00, click the slider, then use the right arrow key to increase the value in steps of 0.25 to 2.00.

Look at all the numbers. I assume your prediction came very close to what they show.

This way of computing the steady state of a control system always works if there is a steady state (if there isn't, the control system design is probably defective). The method has many fancy names in mathematics: it's call Newton's Method (used for computing square and cube roots and roots of equations). It's called a "relaxation method." It's called a "Method of Steep Descent." It's called, as we use it here, "simulation." It is a way of numerically solving algebraic and differential equations both linear and nonlinear, with and without delays, limits, dead zones, and variations in parameters—all without any of the usual manipulations of analytical mathematics. That is one reason I like simulations; I never was adept with complicated differential equations. Simulation is extremely powerful, because it can solve equations when there is no known analytical method for doing so, not just when the solver is less than adept.

There's another thing going on here with considerable mathematical implications. This system is calculating the inverse kinematics and dynamics of the control process without actually calculating them. It is producing just the amount and manner of change of the error signal that is required in order to produce the right output to make the perception produced by the input quantity come smoothly to a predetermined state and remain there despite disturbances. Does the output gain factor fall to half its former value? This system "computes" that it must increase the error signal by a factor of two to go on achieving the same result. Does the environmental feedback factor double? The system computes that the output quantity, and thus the error signal, must drop to half its previous value. Does a disturbance of such and such a magnitude appear? The system computes how much to adjust the error signal to produce enough output to oppose the disturbance. The result is exactly as if something in this system were analyzing the output function and the environment, taking the inverses of the describing equations, and calculating how big the error signal, the signal driving the actuators, must be in order to keep producing a predetermined result. And then, of course, it's just as if the result of the analysis is used to produce exactly the required amount of error signal (within one percent, in the present case).

But there is no such entity doing any such analyses or computations. None of that machinery is needed in this kind of control system. The negative feedback control loop does what is needed automatically without ever taking the inverse of any mathematical form or analyzing the environment in any way.[20]

A couple of chapters farther on, we will see how this same kind of system, with only a small complication, can do (without actually doing it) the same apparent computation of inverse kinematics and dynamics in a far more complex environment. Negative feedback control systems are powerful organizations with a simple structure—just what we would expect to arise through evolution: powerful, to survive selection pressures; simple, to have any chance of evolving through random mutations and selective retention of the results.

As a check on your understanding, you can now try the backward-reasoning method for a change in the environmental feedback factor (the answer was given a few paragraphs above as a question and answer). When you've finished that, you can try one more thing.

Positive feedback.

The environmental feedback slider allows you to change the feedback factor to a negative number. When you do, the product of all the

20 There is a prominent school of thought in psychology and neurology in which it is assumed that "motor control" is produced exactly in the analyze-compute-act manner just rejected. For some reason, it is also assumed that systems that do it that way would be simpler and faster than negative feedback control systems. I suspect that the reason is simply that the properties of negative feedback control systems have not been understood.

signs encountered in one trip around the loop becomes positive. Before that change, there was a negative sign where the perceptual signal enters the comparator; all other signs were positive so the loop gain, the product of all these factors, was negative. That produced negative feedback. By changing the environmental feedback factor to a negative number you make the loop gain positive, so we have positive feedback.

- Try a feedback factor of -0.25 and see what happens.

What happens is that all the numbers go to extreme values and stay there. (If nothing happens, move the disturbance slider slightly.) The program limits those values so that when you play around too enthusiastically with the parameters, you won't always have to restart the program (or your computer) after it halts on an arithmetic overflow. If you set the feedback factor positive again, restoring negative feedback, the control system will recover.

The terms "negative feedback" and "positive feedback" have acquired popular meanings based on a method of definition that my late wife Mary Andrews Powers called "grooving on the sounds of the words" (rather than bothering to find out what they meant to those who invented them). The popular meanings have nothing to do with their meaning in control theory. Negative feedback is always good, positive feedback is always bad, for control systems. There was a time, sixty or more years ago, when vacuum tubes didn't have very high amplification factors, so engineers thought of using a delicately adjusted small amount of positive (or "regenerative") feedback as a cheap way to increase the ability to amplify signals, even in control systems. Control systems actually appeared in which positive feedback played a part. However, it was realized that positive feedback, in order to be used at all, had to be very stable, since only a small change in vacuum-tube characteristics would produce a very large change in the effective amplification, with the upward direction of change leading to disaster and the downward direction to loss of performance. Also, positive feedback greatly amplified noise in the system and made performance highly dependent on exact adjustment of system parameters and on changes in the local environment, a result just opposite to the main virtues of negative feedback systems. So those drawbacks limited the lifetime of that idea. Unfortunately, they didn't limit the lifetime of the idea among non-engineers that positive feedback has a place in control-system models.[21]

There is still a small niche for positive feedback. It is a good way to model processes that, once started, tend to go spontaneously to a limit and then collapse, as in sneezing or even more pleasurable experiences. But such phenomena make up only an occasional part of human behavior and experience, at not a very high level of organization.

A Stimulus-Response System

The control system can easily be changed into a stimulus-response system just by altering some of the parameters.

- First, set the reference signal to zero (you'll soon see why).
  Then, of course, we eliminate the feedback path:

- Set the feedback function's gain to 0.

To compensate for the loss of negative feedback we then have to reduce the input gain to 0.01 or the output gain to 1.00, or some combination of the two that has the same effect.

- Set the disturbance to 15.

You will probably want to set the output time constant to its minimum, 1 second, to avoid waiting five minutes while the output quantity slowly approaches its final value. You will now see that the output quantity becomes exactly -15—precisely equal and opposite to the disturbance.

This precision is somewhat futile, however, because now the former controlled variable, the input quantity, simply changes as the

21 The then president of the American Society for Cybernetics once assured me that positive feedback was good because it told the control system, "Keep on doing what you're doing."

disturbance changes: all control over it has been lost. This system now produces an output that is a faithful response to its input, which might be nice for someone out in the environment who wants to see a specific amount of output response, but it has no relevance to the behaving system since there is no feedback effect from the action. Any disturbing influence applied to the input will simply disturb it.

The effect of changing the reference signal now depends on whether you reduced the input gain or the output gain. If you reduced the input gain to 0.01, raising the reference level from 0 to 15 units brings the output to about 1500 units (the disturbance has a comparatively minute effect). If you reduced the output gain to 1.00 instead, raising the reference signal to 15 units with the disturbance set to zero produces 15 units of output, while changing the disturbance by 15 units has an equal effect on the output in the opposite direction.

In the world of stimulus and response, reference signals are not allowed—they imply that the system is capable of producing spontaneous, or as the behaviorist B. F. Skinner applied his spin, "capricious," behavior. Since that would mean that one could not reliably control the responses (actions, outputs) of organisms by manipulating stimuli (disturbances), and since a non-negotiable assumption was that environmental stimuli cause behavior, the concept of a reference signal was simply omitted from S-R theory. When apparently spontaneous actions were observed, they were attributed to unobserved stimuli. Since there are a lot of those, it was not hard to explain all behavior with this theory.

The Time Plots

Finally, it's time to talk about the graphical plots that you turn on by clicking the GRAPHS ON button. To the right of the plotting area, which behaves like a strip-chart recorder on a heart monitoring machine (very nice, Dr. Abbott), are colored-coded boxes which you can check and uncheck by clicking on them. The names of variables are printed by each box. When you check any of the boxes, that variable immediately starts to be plotted, positive upward, negative downward from the centerline of the plotting area. For each instant of time, the checked variables are plotted in order from top to bottom. Plots lower on the list are plotted last, over higher ones in the same screen location: if two of them have the same numerical value only the one listed last will be seen. If a variable seems not to have been plotted even though it's checked, its trace is exactly behind another trace. The numbers in the diagram show the values of the plotted variables, while the plots let you look at temporal relationships among the variables. A slider on the right lets you adjust the vertical scale of the plots.

As the plot unreels to the left, tick marks appear at intervals of one second. If you click on the PAUSE button, just above the plotting area, the plot will continue to the next tick mark and stop there. This helps to see the timing of changes. Clicking the button again (it now says "Resume") will start the plot going again. You can change parameters while the plot is paused or while it is progressing.

Near the top left of the screen is a **Slow** motion radio button. Clicking one setting changes the time scale of the plot by a factor of 10 and unclicks the other setting. While in slow motion, each tick mark represents the passage of 0.1 second of time instead of 1 second. You can switch back and forth from normal to slow motion at any time. When you change a slider's setting, you are working in either normal or slow motion: the affected variable always changes right along with the slider movement, so in slow motion you are moving the sliders ten times as fast in simulated control system time.

There is a way to make instantaneous changes in parameter settings. Click the PAUSE button, then change a slider setting (or more than one), then click resume which has replaced the Pause legend. The program will think that any changed variable has changed in the time of one iteration, 1/60 second, which is instantly in the world of this model. The PAUSE button is useful for making exact changes in parameters as well as for halting the action so you can look at the numbers in the diagram.

- As an introduction to using the plots, set the output gain factor to 30 and the output time constant to 8.
- The left and right arrow keys will help with fine adjustments.
- Set the reference signal to zero and the disturbance to -15.
- Wait until the traces have leveled out, and click PAUSE to stop at the next tick.
- Click the **Slow** radio button in the upper left corner so we will see what happens next in slow motion.
- Now change the disturbance to +15.
- Click RESUME. The plot starts.
- When the initial jump in the red line is almost at the left margin, click PAUSE again.

What you see will look something like Figure 3-3:

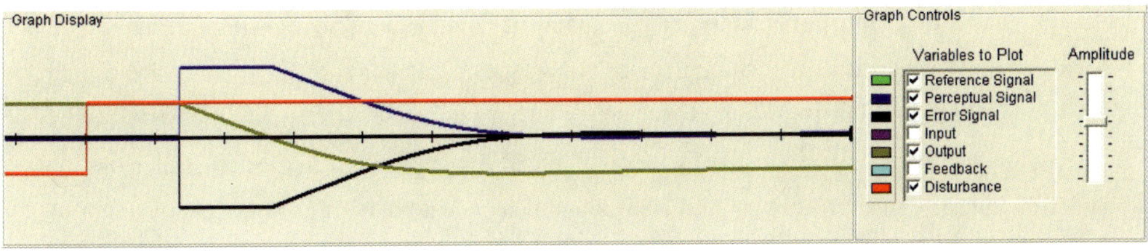

Figure 3-3. Disturbance (red) jumps from -15 to 15. After a delay of 133 milliseconds, the perceptual signal (blue) jumps positive, the error signal (black) jumps negative, and the output (olive) starts to go in the negative direction. The feedback effect of the output change on the perceptual and error signals is delayed another 133 milliseconds, then those variables begin to change toward zero along a smooth curve. Time-ticks are now 0.1 second—100 milliseconds—apart.

This is how you can use the plots to explore the relationships among variables and see how their simultaneous changes come about. Even though there are time delays, every variable always has a value and is always changing until all variables stop changing. Ordinary causal or sequential thinking will not be adequate for understanding how this simple unit of organization works, or to predict how it will behave. You can't build up a picture of the whole system just by examining each part of it in detail—the causal relationships you see in the details are not the causal relationships that operate in the system as a whole, as we saw when changing the gain factors. A lesson, you might say, for even the most accomplished of us in other fields. The forest does not look like the trees.

There are many experiments you can conduct with this Live Block Diagram. As you do so, you will start to build up an intuitive feel for how a negative feedback control system works, as an antidote to all the years through which your intuition about stimulus-response systems was built up by your education.

What you can learn from experimenting with this simplified, idealized, and isolated control system simulation will prepare you to begin looking with new eyes at the real system it models. This applies if you study human behavior as a whole, if you are a neurologist tracing circuits in the brain, or if you are a biochemist tracking genes, signaling molecules, and cell functions. Enormous amounts of data have been accumulated about isolated localized causal relationships in behavior, in the body, in the organs, in the cells, in the genome. But almost nothing has been accomplished in the way of understanding the whole system, because the kind of causal thinking that is appropriate for studying localized relationships breaks down when those relationships become organized into closed loops, into control systems. The human organism, and every living system, is a collection of control systems at every level of organization. That is the hypothesis being expressed at great length in this book.

With that sermon ended, we can now return to the subject of applying models to behavior. In the next chapter we will look at simple tracking behavior, a subject to which control theory has been applied—and misapplied—since the late 1940s. We will use the model in this chapter, adjusting its parameters until it can reproduce the behavior that you show while making a pointer track a moving target. One chapter later we will look at a slightly more complex model, a two-level model, which also fits that behavior, but which can, in addition, show apparent adaptation to changes in the properties of its environment—an adaptation that is, I am afraid, yet another illusion generated by the strange properties of negative feedback control systems. Control systems can seem to adapt themselves to a changing environment when they are not changing their characteristics at all.

Chapter 4: Tracking

If Chapter 3 succeeded in its aims, you now have a considerably enhanced idea of what a negative feedback control system is and how it works (some readers, of course, already knew). This is only a building-block; a complete theory of behavior would use many of these control systems operating in parallel and at many levels of organization. But even one building-block can be applied to isolated behaviors of many kinds from the simple to the complex, in each case suggesting how some low or high level of control might be organized. The object of this chapter is to see how well the model in Chapter 3 can describe and predict what we observe in a simple tracking task.

SIMULATION AS A SCIENTIFIC METHOD

The strategy, which we saw briefly in Chapter 2, involves constructing a working model and using it to simulate human behavior in a set task. The model, when expressed as a simulation, is the equivalent of a detailed and quantitative prediction of behavior that can be compared with real behavior.

The two main aspects of a model are its architecture and its realization. The architecture is the block diagram in Figure 3-1 in Chapter 3, without any mathematics—just labels for the parts, the connections among parts, and general descriptions of what the parts are and do. A block diagram is really a whole family of models that can behave entirely differently depending on which realization of the model you use. The realization of the architecture consists of the mathematical forms that go into each function box, and most particularly the values of the parameters or constants in the equations. These are what make the model real. Our model will contain adjustable parameters, nearly the same parameters that were being adjusted with sliders in the Live Block Diagram of Chapter 3. We saw, for example, that all you have to do is change one number in the model from positive to negative to make the model behave totally differently—literally, to stop controlling. There are numerous ways of adjusting the parameters that will produce behavior completely unlike what a real person would show, far more ways than there are to produce realistic behavior.

That, in fact, is the feature of this approach to modeling that we can count on to give not just a model that generally reminds us of real behavior, but a model that reproduces real behavior in some detail, and that can even predict real behavior under changed circumstances with almost as great fidelity. Generally, there is just one set of values of the parameters that will make the model's behavior match the real behavior with the least possible difference. That set of values will make the model fit not only the behavior that was just observed, but the behavior that will be observed in the future, even under changes in the experimental conditions. That fact puts this modeling business into a new perspective. It suggests that we can use this approach to *measure* parameters that characterize a real human control system, parameters that are *properties of the behaving system* and that remain reasonably stable or change only with marked changes in conditions.

We can expect this sort of fit of model to real behavior, and this sort of predictive power, only for well-learned behaviors under normal conditions—no starvation, sleep deprivation, drugs, surprises, or distractions. The effects of such things can be studied, but first we need to establish a baseline: a model that fits ordinary well-learned behavior without any extreme circumstances to complicate matters. I stress well-learned—an important assumption made in matching the model to real behavior is that the parameters in the real system are the same throughout an experimental run. A simple model does not venture to include mathematical descriptions of how the parameters change during learning. In the future that will be an interesting subject of study.

A Tracking and Task Mode

Make sure your demo CD is up and running.

1. Open Demo 4—TrackAnalyze

2. Click the COLLECT DATA button.

3. In the Track Analyze pop-up window, select the file called default.txt.

4. Click SAVE.

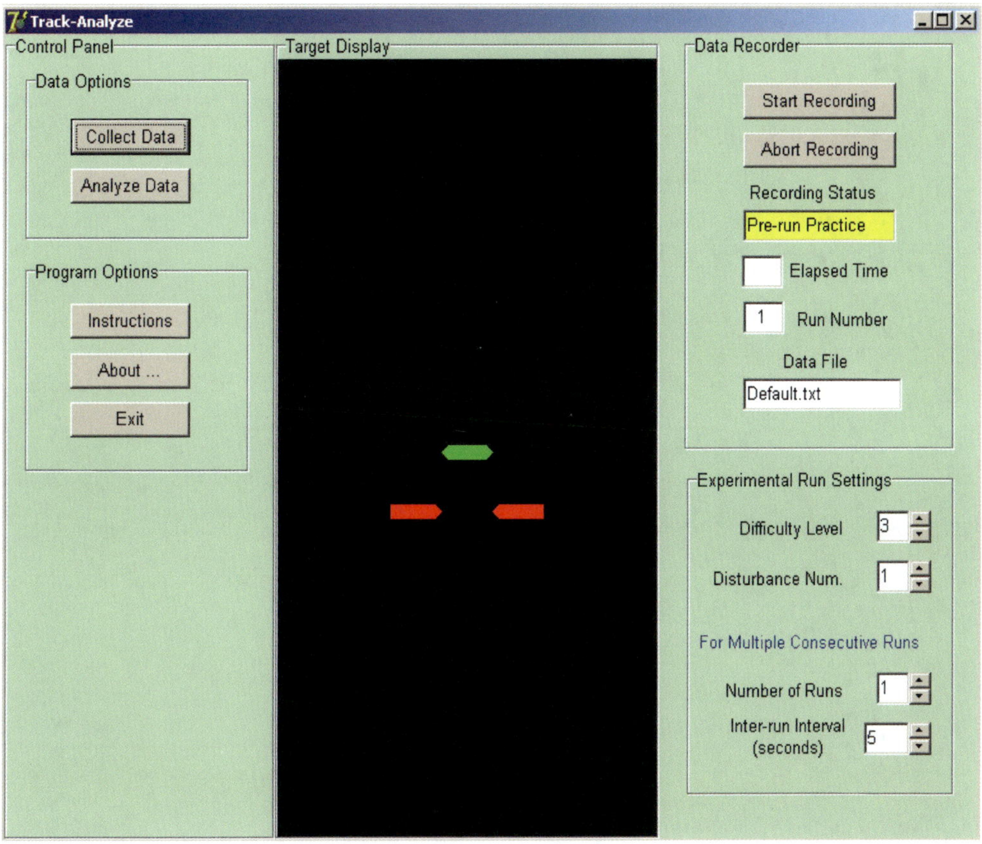

Figure 4-1. Tracking program after a file name has been selected or entered.

Note that reminders and tips are available through the INSTRUCTIONS button, including how to create a new set of random disturbance tables (delete the old one). Two red bars move together, up and down, in the black frame. A green bar with pointed ends is stationary on the screen until you move the mouse pointer into the frame. Then the green bar jumps to the same vertical position in the frame and follows the mouse pointer.

- Keep the mouse pointer on the green bar while you're doing this task (actually, keeping the pointer anywhere inside the frame will do).

Practice holding the green line between the red bars as accurately as you can—take a few minutes for this. This is called *pursuit tracking*—you are making the green line pursue the red target lines. When you feel comfortable with the tracking.

- Start an official run by clicking START RECORDING.

You will have four seconds to move the mouse pointer into the frame to start tracking (you don't need to click anything). If you make too many mistakes or the mouse pointer moves outside the frame so the green bar freezes, just click abort recording and start again. After one minute the run ends and the tracking bars disappear.

- Click ANALYZE DATA to open the Analyze window.

If you're the first to use this program on this computer there will be just one file name in the Data File field; default.txt.

- Click OPEN DATA FILE to prepare default.txt for receiving data.

This is the result:

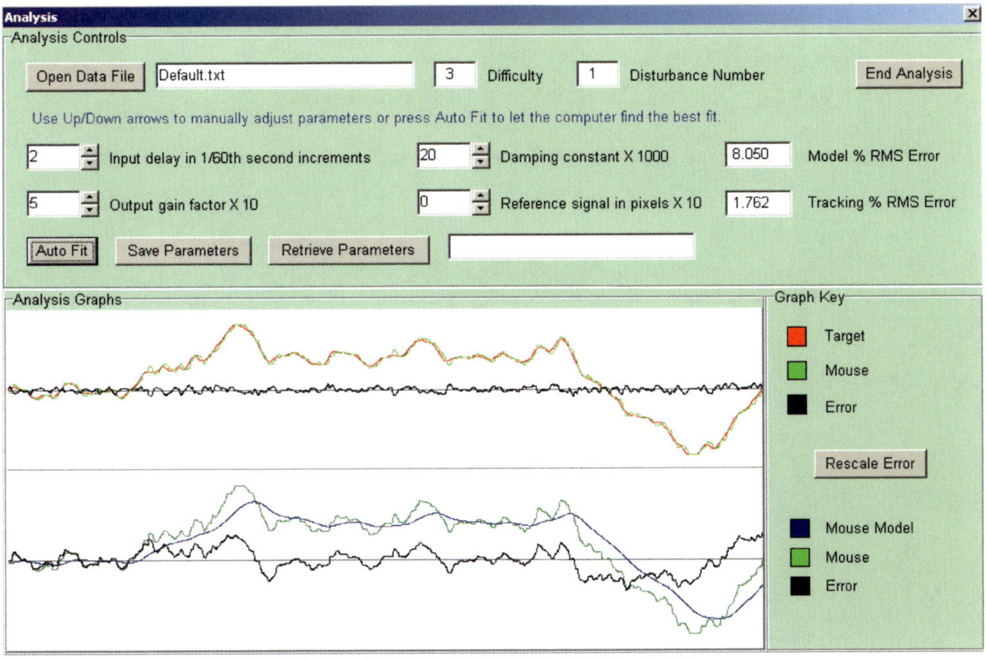

Figure 4-2. Analysis screen - the run data is in top plot, initial rough fit of model mouse positions to real mouse positions is in bottom plot.

The top plot shows the run you just did. The green line shows how the green bar, the cursor, moved. The red line plots the movements of the red target bars. The black line shows the distance from the target to the cursor; a straight grey line indicates zero. Notice the Tracking % RMS Error field which shows the average (root-mean-square) deviation of the mouse from the target as a percentage of the maximum range of target movement.

The bottom plot compares the mouse positions from the data plot with the simulated mouse positions generated using the model of Chapter 3. This plot is actually being retraced 60 times per second, so the model is being put through a simulated one-minute run 60 times per second, or 3600 times the actual speed. The model data are in blue, the real data in green. A black trace now shows the difference between model and real data, and a straight grey line indicates zero.

Check the Model % RMS Error field. This is the RMS deviation of the model's mouse behavior from the real behavior, again as a percentage of the maximum range of target movement. When errors are small, to get a better look at them, click RESCALE ERROR to expand both error traces. Click again to shrink them.

Above the plots there are four spinners, boxes containing a number and two triangles you can click on, one to increase and the other to decrease the numbers (the only way to change them). The top left spinner sets the input delay in 1/60 second increments; the one below it sets the output gain factor (times 10) in pixels per second per pixel of error; the top right one sets the "damping" (leakage) constant (times 1000) in fraction lost per second, and the bottom right one sets the reference signal's value (times 10) in pixels. These are the four adjustable parameters in the model. To read the actual values, mentally divide by 10 or 1000 as appropriate. By adjusting these parameters you can minimize the difference between the model's simulated mouse behavior and the real mouse behavior, the Model % RMS Error. The plots change as you change the parameters.

First, try doing this manually. Adjust the spinners to get the smallest Model % RMS Error that you can obtain. The green trace shows the real mouse behavior and the blue trace shows the model's mouse behavior. Try it in this order: (1) output gain factor, (2) reference signal, (3) damping constant, (4) delay. Adjust each spinner to get a minimum reading of the upper difference number, which shows how well the model matches the real mouse behavior.

Keep cycling through the adjustments until you can't get the difference any smaller. You will see when you're finished that while the model's mouse movements are not exactly the same as the real mouse movements, they are rather remarkably close. You will find that making any parameter either a little bit greater or a little bit less than its final value will make the difference bigger—not by much, but quite certainly. The model is optimized when the model % RMS Error number is as small as possible. It doesn't prove that there is no other combination of values that would give an even smaller difference, but in fact, for this model, that is quite unlikely. You can look for other combinations with significantly different parameter values, if you wish, but I predict that you will give up before you find a better one. After you've optimized the model manually once, you will very likely prefer using the AUTO FIT button. Figure 4-3 shows the result of using auto fit on the data from Figure 4-2.

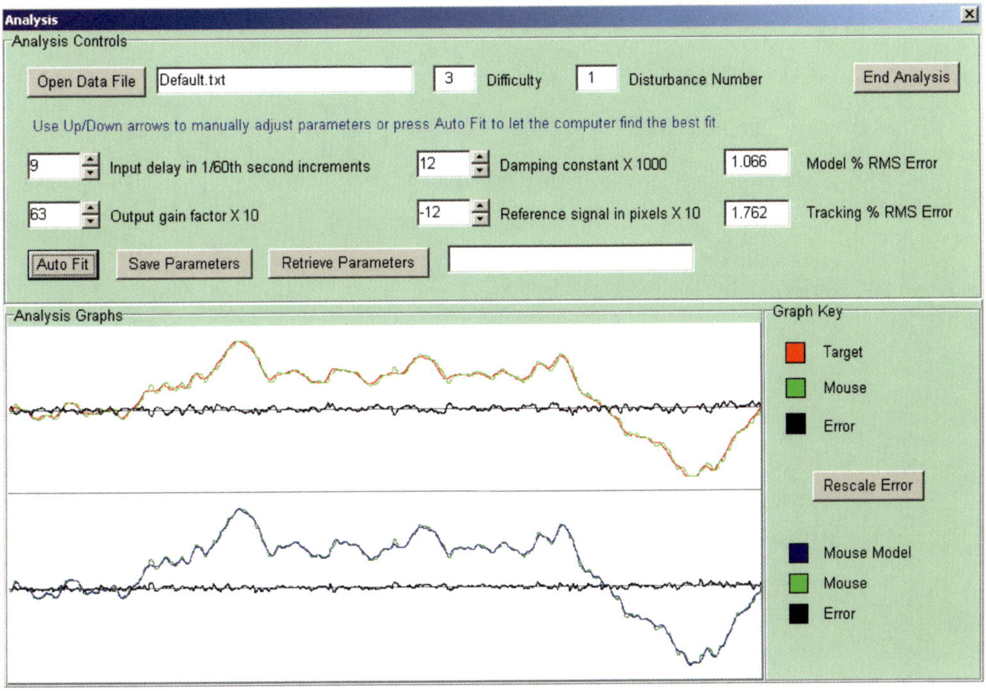

Figure 4-3. Fit now shown in the lower plot after using the auto fit button.

Making Predictions

What we just did was to fit a model with a particular architecture to a real data record by adjusting the parameters of the model to find its best possible realization. Under the assumption that this architecture is correct—that it adequately represents the actual architecture of the human system—the parameter values we end up with can be taken as *measurements* of the parameters in the real control system. The question now is whether these so-called measurements deserve that name.

If you're really measuring properties of the human being who generated the data, then you should find the same values of parameters twice in a row.

- Before ending the analysis, click SAVE PARAMETERS.

This creates a file called default.par which contains the four parameters and the two fit errors, overwriting the existing file if one exists.

- Now use the END ANALYSIS button to complete the task.

Repeat exactly the same experiment, leaving the difficulty set to 3. However, select a different disturbance number (2 to 10)—the parameters should not change just because the environment behaves a little differently. Click collect data, assign a new file name (say, NewRun), click open data file, then start recording and do the run. (If you forget to start recording, you can go on forever in practice mode without accomplishing anything.)

- After the run, click ANALYZE DATA and pick the appropriate data file.

You will see the mouse behavior for you and for the model, with the parameters where they were left after the previous analysis. Look at the number for the model fit error. It will be almost as low as it was after you fit the model to the previous data. This is now a prediction of your behavior, and you can see how much it differs from reality.

Wait a few hours, or a day or two, or a week, and see how well the original set of parameters accounts for the data. One researcher (Bourbon, 1998) has published a simpler version of this test for consistency of the parameters over a span of five years, with similar results. It does seem that this method is measuring something real.

This program can be used to do a series of runs with a pause between them. You can leave the difficulty and disturbance number the same to explore learning in a naive subject, or you can specify a longer pause to allow changing one or both numbers between trials. The program will give the files containing the results of the trials the same name that you designate, with the second and following names containing _b, _c and so on.

SUMMARY

The aim in this chapter was to show how a working model can be constructed that mimics a simple control behavior, and how adjusting the parameters of this model can result in very good predictions of behavior under new conditions. This is much like the normal hypothetico-deductive form of experimental investigations, with the difference that the hypotheses take the form of a working model, and the predictions from the model appear as the quantitative behavior of a simulation.

The control-system model is strongly challenged by this approach. We are not asking that the model behave in a way that resembles real behavior in a general way, most of the time, in a majority of subjects. Rather, we are asking that the model quantitatively fit and predict real behavior within narrow limits every time it is tested, in every subject. We can't claim "every time" yet because that would require living a rather long time, but this sort of fit of theory to measurement is found, so far, in all subjects tested, given a few practice runs.

The particular behavior involved isn't of much general importance (though it can be used in any case where a motor action is used to control a perceptual variable). But it provides more reason to think that the control-system model is pertinent to human organization; it is certainly pertinent to explaining tracking behavior, and provides some reason to look further for more examples of control behavior that seem to involve similar principles.

Tracking is just one form of control; it's done by controlling a visual relationship, one part of which varies independently (the target) and the other part of which is affected by the participant's actions. You already know that there are other forms that control behavior can take because in Chapter 2 you experienced yourself controlling orientation, position, or shape of an object, depending on which aspect you selected to control. We are almost ready to broaden the view to see how this model might apply in a wider variety of circumstances, but first we are going to bring up the subject of adaptive control systems. In the next chapter we will see that what appears to be adaptation is not a reliable indicator that anything about a control system has changed. A control system with unchanging properties can continue to control well over a wide range of environmental conditions.

Reference

Bourbon, W. T. , Copeland, K. E., Dyer, V. R., Harman, W. K., & Mosely, B. L., 1990). On the accuracy and reliability of predictions by control-system theory. *Perceptual and Motor Skills*, 71, 1990, 1331-1338.

FOR THE MATHEMATICALLY COURAGEOUS

For this discussion, Figure 3-1 in Chapter 3 represents the experimental situation. Let T stand for target position, which is continuously changing in a smooth but random way. It plays the part of the disturbance in the Live Block Diagram. Let C stand for cursor position, positive being above the centerline. The vertical distance of the cursor from the target is the input quantity q_i that is being controlled. A positive distance means that the cursor is above the target, so you can write the equation for the input quantity:

$$q_i = C - T.$$

All of the variables being defined here change with time, which is a variable indicated by a lower-case t. Strictly speaking we should include that fact in referring to variables: we should say C(t) and T(t) instead of just C and T. However, if you keep that time-index in mind, we can write equations that are less cluttered-looking.

This is the environmental part of the model; all these quantities are observable in the world outside the behaving system so there is no doubt about this part of the model.

Now we turn to the model of the behaving person, which is like the part of the Live Block Diagram of Chapter 3 above the System-Environment Boundary.

As before, the perceptual signal represents the input quantity with a delay and a scaling factor of 1, so we can write (the time indices are necessary now):

$$p(t) := q_i(t\text{-tau}),$$

tau being the amount of time delay.[22] Note the colon-equal sign showing replacement rather than equality.

22 The notation p(t) means the magnitude of the perceptual signal as it exists at any clock-time t. $q_i(t - \text{tau})$ means the input quantity as it existed tau seconds prior to time t—that is, at time t - tau. This is how transport lags are expressed mathematically.

The error signal is, as always:

$$e(t) := r(t) - p(t),$$

where r is the reference signal setting. The reference signal specifies the desired distance between the target and the cursor.

The output function contains a time integrator, meaning that its output q_o has a value that is cumulative over time. (This will be explained in more detail later.) The change in output quantity during one iteration of the program, 1/60 of a second, is symbolized dq_o, where the d is the notation in the calculus for a change in the value of the symbol that follows it, q_o. The basic form of the output function creates a change in the value of q_o every 60th of a second that is proportional to the amount of error signal. K_o indicates the constant of proportionality, often referred to as the gain of the output function:

$$dq_o := K_o * e$$

The asterisk here indicates multiplication explicitly and the colon-equal sign indicates replacement rather than equality, as explained in Chapter 1. Last reminder.

This equation shows the change of q_o per second, but we need to know how much change there is in one iteration, which lasts 1/60 second. Using dt to represent the elapsed time per iteration (so we can change the iteration time if we wish just by setting the value of dt to a different number at the start of the program), we end up with:

$$dq_o := K_o * e * dt$$

In this model we add a small embellishment to this output integrator: a bit of leakage, rather like the hole in the bucket encountered in Chapter 1. On every iteration, the output as, it now stands, changes by an amount equal to K_o*e*dt, and if e falls to zero, the output remains constant at its last value. To make the output change in the other direction, e has to go negative. But in real devices and real nervous systems, the output of integrator is always tending to decline toward zero, like the water level in that leaky bucket. The amount of change in the value of q_o in one second due to leakage is simply $-K_d*q_o$. The negative sign says there is a decline in the value. The constant K_d, if large, says that the hole in the bottom is large and the leakage is rapid. If K_d is set to zero, someone has patched the hole. K_d*q_o says that the leakage per second gets larger if either K_d is larger or q_o is larger. For a constant error signal, the output will rise until the negative effect of the leak rate equals the positive effect of the error signal. K_d is called the *damping constant* because of its effects on dynamic changes—not because I am using a rather damp analogy here. We can say, finally, that:

$$dq_o := (K_o*e - K_d*q_o)*dt$$

The expression on the left, dq_o, means "the change in q_o." This value will be added to the previous value of q_o to obtain the new value, in each 60th of a second. That is, $q_o(new) := q_o(old) + dq_o$. That is called "Euler integration," the simplest form of integration which is nothing but a summation of values over time. We have to bring in the time indices to show that the "new" value occurs one iteration after the old value, after time advances by dt, or 1/60 second:

New value = Old value + [Change in value]

$$q_o(t+dt) : = dq_o(t) + [K_o*e(t) - K_d*q_o(t)]*dt$$

The integration and the leakage factor together are equivalent to an amplifier with a time constant. If an error signal suddenly appears and remains constant, a change in the output quantity during each interval dt will begin. But as time passes, the size of the output quantity increases so the leakage also increases. Eventually the amount of leakage in each dt will equal the amount of accumulation, and the output quantity will level off. The time constant you saw in the Live Block Diagram is, in fact, equal to K_o/K_d.

Summing up, our model is now represented by the following input-output relationships describing each function in the control loop (time indices shown only where necessary). Since the cursor position is directly affected by the output quantity or mouse position, substitute C for q_o.

1. qi = C - T [Environmental effects]

2. p(t) = qi(t - tau) [Input function]

3. e = r - p [Comparator]

4. C(t+dt) = C(t) + [Ko*e(t) - Kd*C(t)]*dt [Output function]

Again, t is time, tau is lag, and dt is the physical time interval represented by one iteration of all four steps in the program. This is exactly the model shown in Chapter 3's Live Block Diagram, with the target position T playing the part of a disturbance. The gain is equivalent to Ko/Kd, and the time constant is 1/Kd in seconds.

When Kd approaches zero, the output function becomes indistinguishable from a pure integrator and the time constant of Chapter 3 goes to infinity. A 30-second time constant is equivalent to a damping factor of 0.033.

Chapter 5: Nonadaptive Adaptive Control

Among the many intriguing properties of negative feedback control systems is their apparent ability to adapt to changing environments without altering their characteristics at all. The point of this chapter isn't that a control theory of human beings can ignore learning and adaptation, but that negative feedback control can work across a wide range of conditions without the degree of adaptation that in the past has been assumed necessary. Negative feedback control systems can function without needing changes in their properties to meet every little variation in the environment, the way other kinds of behavioral models need to change—even some models that are called control-system models. You have seen a hint of this in exploring the Live Block Diagram.

We will have our first look here at a multileveled control system, a very simple one with only a single control system at each of two levels. The lower level controls a perception of velocity by altering the force applied to a movable mass in the environment. The second controls a perception of the position of the mass by mean of varying the reference signal sent to the velocity control system. This arrangement actually approximates the control systems that exist in the vertebrate spinal cord, where *phasic* stretch receptors in the muscles report the rate of lengthening of muscle fibers, approximately the rate of change of joint angle, and *static* stretch receptors report the lengths of the muscles, an approximate measure of joint angle itself. There is actually a third level in the spinal control systems.[23] lower than the other two, in which signals sent to the muscles control perceptions of tendon forces (proportional to acceleration), but we will merge that level into the velocity-control level in this model.

The point to be made in this chapter is that when we apply standard methods of measuring control performance to this two-level system, we will get a very clear indication that the system is altering its own dynamic characteristics when the characteristics of the load it is moving change. When a load consisting of a very small mass is given a much larger mass, the model modifies its application of force to overcome the effect of the increased mass, so the mass moves nearly the way it moved before. When the load begins to act as if it is moving in a thick, viscous syrup, the model pushes harder in proportion to the velocity. When a spring is added that tends to move the load toward one central position, the controller adjusts its action to compensate for that, too. And in fact the model is not changing itself at all—it doesn't have any ability to alter its own characteristics (though you can alter two of them yourself, in the simulation). Since our model has known characteristics, we can state quite positively that its characteristics are not changing even though they appear to be doing so.

This will come as a surprise to many engineering psychologists, because about 40 years ago, several of their ancestors (McRuer and Jex 1967) determined experimentally that human control characteristics *do change* when the characteristics of the load change

23 What I am calling spinal control systems are what others call, more traditionally, spinal reflexes. In the real system, the arrangements are more compact than they are in this model, although the net result is mathematically the same. In fact, one could argue that the velocity control system is at a level higher than the position control system, or that they work at the same level. The interpretation I use is reasonable, but at this point (this is only the 21st Century) we remain ignorant of many things.

from leading to proportional to lagging. There is no implication here that their measurements were in error; they were correct, but what they measured was probably an illusion.

This is an important illusion from several standpoints. First, it shows that negative feedback control is much more powerful than other proposed architectures as a behavioral model because even without adaptation it is robust against many kinds of changes in the environment. And second, it implies that the requirements for evolving successful negative feedback control systems through natural selection are far more relaxed than they are for evolving other kinds of systems—systems that can behave properly only after analyzing changes in environmental properties, for example. That gives the negative feedback control system a large competitive advantage in the struggle for survival (the real struggle, and the academic one). Natural selection becomes far more tenable as a theory of evolution if we assume that what evolves are negative feedback control systems, prevailing over other kinds of system that may arise.

For readers familiar with Bode plots and frequency-domain analysis, the results of the simulation in this chapter can be summed up very briefly. When the load is a mass, the control system's output amplitude for a sine-wave input (error) amplitude (its frequency response) increases at 20 db per decade. When the load is a pure viscous resistance, the frequency response is flat. And when the load is an elastic restoring force, the response decreases at 20 db per decade. This is observed when real people control loads with variable characteristics, which is what convinced McRuer and Jex and many of their intellectual descendants that the human system was adapting itself in response to the changing load characteristics. But, as the character Sportin' Life remarked a few decades before McRuer and Jex, "It ain't necessarily so."

THE MODEL OF THE ENVIRONMENT

Before we get into the control system part, let's look at the model of the load moved by the control system, which will be a mass (adjustable) on a spring (adjustable) with damping (also adjustable).

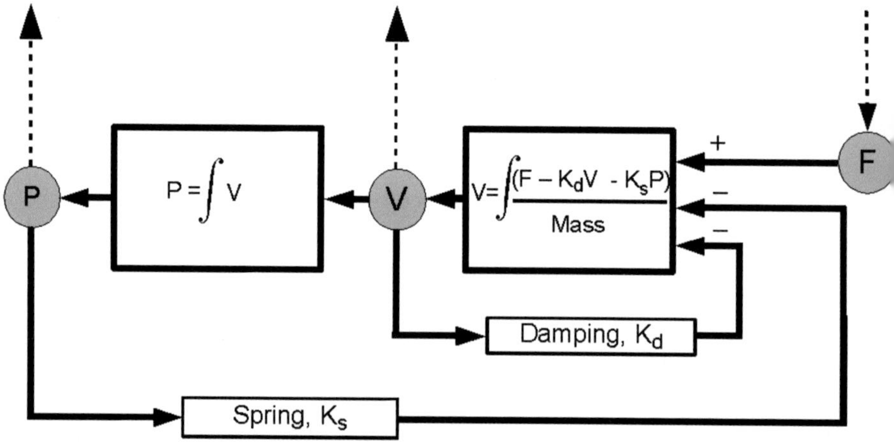

Figure 5-1. Analog computing setup for simulating a mass on a spring with damping. This load corresponds to the environmental feedback function in the Live Block Diagram in Chapter 3 (see Figure 3-1).

The model of the load contains two consecutive integrations and two feedback paths. This is how a mass on a spring, or more generally a second-order differential equation, is set up on an analog computer.

The first integrator receives an input labeled F for force from the output function of the control system. The integrator's output will increase at a constant rate when the total force applied at its input (the sum of the three forces shown) is held steady at one level (greater than zero). The output of the first integrator block is labeled V for velocity. Isaac Newton would approve, because we have just modeled one of his three laws of motion: the velocity of a mass changes at a rate proportional to the net force applied to it. A computed integrator creates that same relationship between its input and output. The constant associated with the first integrator expresses the ratio of change in velocity to applied force. It is numerically equal to the reciprocal of mass. The program allows you to change the mass, and the result is applied in this part of the model.

The second integrator receives the velocity as an input, and generates an output that is labeled P for position. If the velocity is constant at some value other than zero, the position will be changing at a constant rate. That makes sense: if a speedometer reading is steady and not zero, the car is changing its position at some constant rate, so many miles of positional change in each hour. This is just a basic law of our particular universe, for velocities well below the velocity of light.

This mass is effectively mounted on a spring with one end anchored to a fixed point at position zero. Since a zero-length spring is hard to imagine, you can think of the mass as suspended between two long springs so without any forces applied the mass sits at the zero point of the coordinate system. If the mass is moved some distance away from zero, one spring will shorten and the other will lengthen, generating a net force that tends to restore the mass to the zero position. The restoring force is represented by the longer of the two feedback paths in Figure 5-1, the one starting at the position label and called *Spring, Ks*. The restoring force acts oppositely to the applied force, as shown by the negative sign on the feedback path where it enters the first integrator along with the force being applied by the control system. The ratio of restoring force to one unit of displacement is called the *spring constant*; symbolized Ks. A large spring constant means a stiff spring that creates a lot of restoring force when stretched to the left or right.

If this load is immersed in a thick liquid, another resistive force develops, this time depending on velocity. The faster the load moves, the more resistive force develops, also opposing the applied force from the control system. That is shown by the feedback pathway coming from the velocity and labeled damping. The greater the damping, the more resistance there is to a given speed of motion. This *damping constant* is the third adjustable property of the load, symbolized Kd.

There is one more feedback force, the reactive inertial force due to acceleration, but in this model we simply show the net applied force as if the reactive force has already been taken into account. That level isn't necessary for making the main point. Inside the integrator boxes are differential equations which say in mathematical notation what has just been said in words. The tall thin curly S is the symbol for integration.

THE HIERARCHICAL CONTROL MODEL

Figure 5-2 shows the hierarchical control model we will use to control the mass-spring load.

The position of the load can be varied by a higher-level system that can adjust the position reference signal. This reference signal acts like a target toward which the control system always tries to change the perceived position of the load. The organization is quite similar to that of a tracking control system that moves a load (the hand and arm operating the pointer) to keep it near a moving target. In fact, the error signal in the higher control system corresponds to what the engineering psychologists of the 1950s called the input to a tracking control system, and the position of the mass corresponds to the pointer that a subject in a tracking task moved to align with the target. The apparent adaptation of this system was found by attaching different kinds of loads to the hand and arm and measuring the relationship between the position error and the force applied to the control stick. We will do that with this model, too.

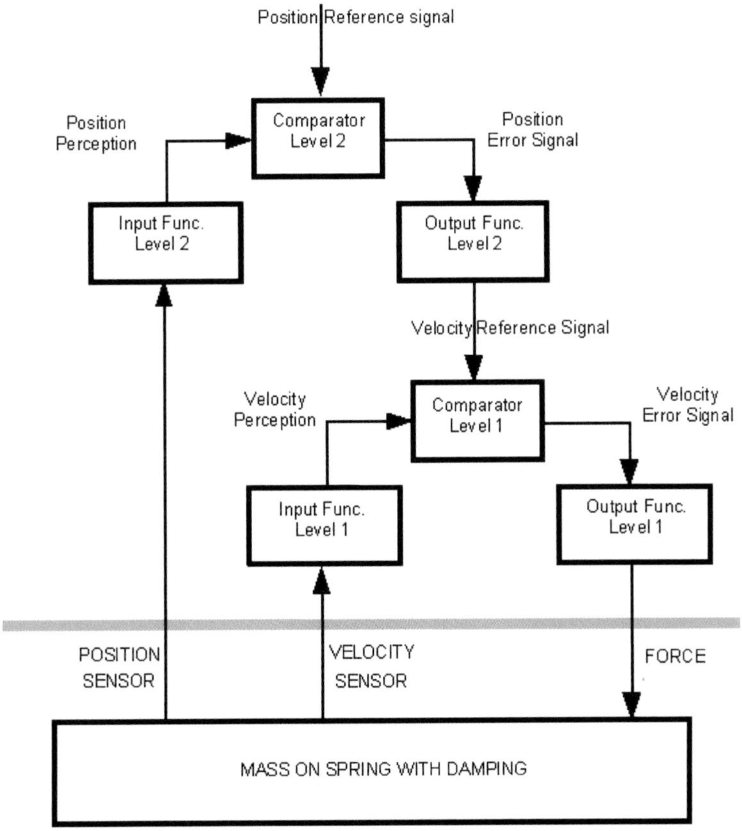

Figure 5-2. A two-level control system controlling a mass on a spring with damping. The lower level senses the velocity of the mass, the higher level senses the position. The position error is amplified in the higher-level output function to become the reference signal specifying the velocity that the lower system is to produce. The lower system compares the sensed velocity to the reference signal, the error signal being amplified by the first-level output function to cause a force to be applied to the mass in the environment (below the gray line). This arrangement could be reduced to a single equivalent system, but doing that would conceal interesting features of this architecture, and would misrepresent any real system actually organized as shown here.

- Time to start the program (Demo 5-1). When you do, you'll see this screen.

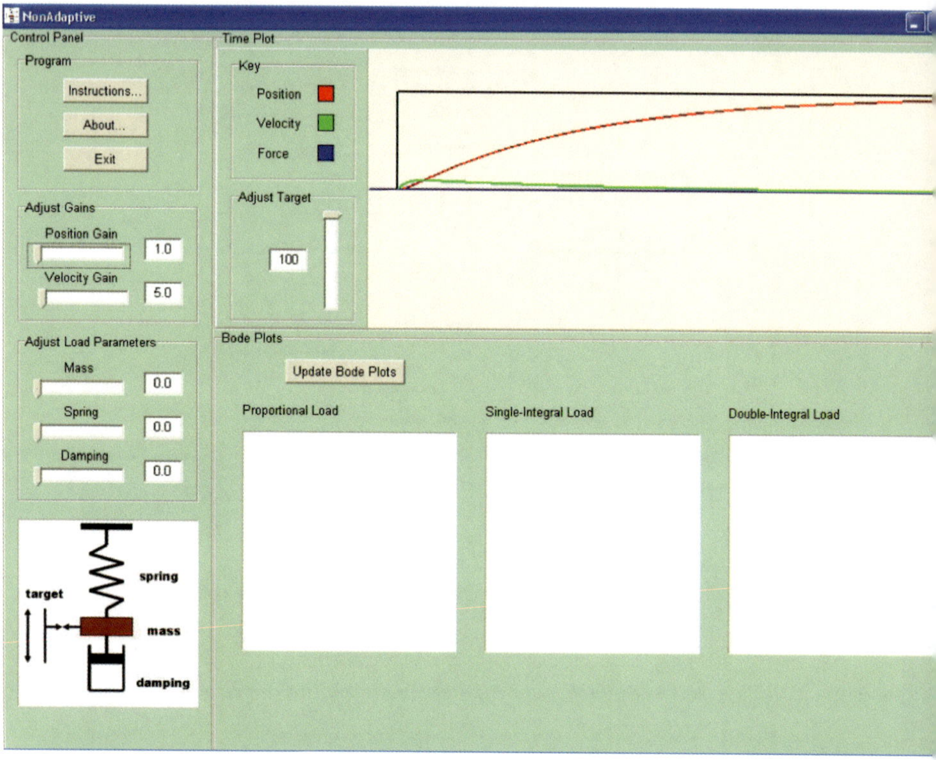

Figure 5-3. Mass on Spring control system. The simulation runs 20 times per second. The mass is started at position zero (left end of horizontal blue line), and the position reference is set to zero. The simulation is started, and after a short delay the position reference signal (black) jumps to a steady value (which can be adjusted using the Adjust Target slider on the left. The plot continues and after several simulated seconds reaches the right side of the screen. The position of the mass, shown by the red line, changes until it matches the reference position. Then everything is reset and the next trace starts.

With the repetitive plot of the behavior of this two-level control system, you can observe the effects of varying the adjustable constants in this system: the output gain factors at the two levels, and the three properties of the load: mass, damping, and spring constant.

Right now the Position and Velocity gain factors (top two sliders) are set near the low ends of their scales. If you move the sliders for each factor to the right to the maximum values, the control of mass position will improve considerably.

- Adjust the two gain factor sliders now.

The sensed position of the mass rises more quickly to the same value as the reference signal just after the value of the reference signal steps upward. By increasing and decreasing the final value of the reference signal using the slider, you can verify that the final position of the mass, after the initial lag, tracks the setting of the reference signal.

With the two gains set to maximum, and the three properties of the load set to their minimum values, the trace looks like this:

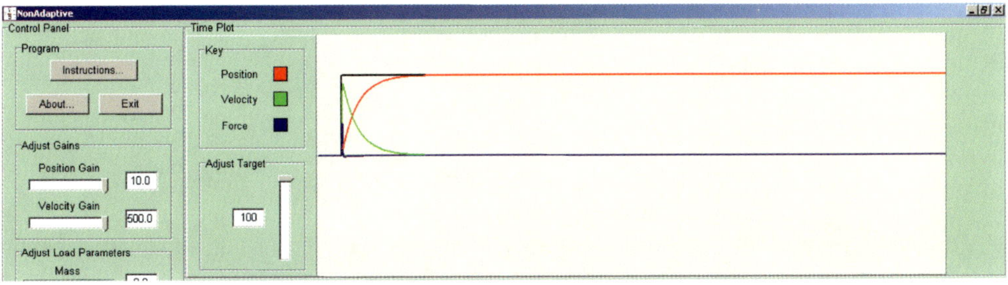

Figure 5-4. Position gain and velocity gain set to maximum values, load constants set to minimum values. The minimum value for the mass is actually 0.2 units, simulating the limb that is moving the mass. A total load mass of zero would be physically impossible. (Only the variable part of the mass from 0 to 20, added to the minimum, is shown in the figure.)

The green line shows the velocity of the mass. It rises very quickly, almost instantly, to a maximum value because the force applied to the mass (shown by the blue line) becomes large at the instant the reference signal jumps from zero to the indicated value. It's hard to see in this plot, but as soon as the velocity becomes very large, the force drops to a small negative value, applying a braking force to the mass. The mass is actually coasting inertially after the initial force impulse, slowing down because of the reversed force. The drop of the green line toward zero shows how the velocity decreases. The negative force gradually decreases, reaching zero just as the mass reaches its final position. The spring and damping constants are zero, so this is a Newtonian mass, moving in a vacuum at a constant speed when the total force acting on it is zero, like a piece of equipment being moved by an astronaut in orbit.

It's important to realize that these force adjustments, while perfectly logical and reasonable, are not generated either by logic or by reason.

They arise naturally out of this two-level arrangement of control systems. This system does not need to be told how much force or how much braking to apply, or just how to reduce the braking so as to stop the mass just as it gets to the right position. That happens completely automatically without any special computations.

- Leave the two gain factors in the control systems set to maximum from here on.

Now we will look at the effect of varying the properties of the load. The most dramatic effect comes from increasing the mass of the load, adding to the 0.2 mass units that are always present.

- Move the mass slider all the way to the right, increasing the total load mass to 20.2 units, or about 100 times the previous value:

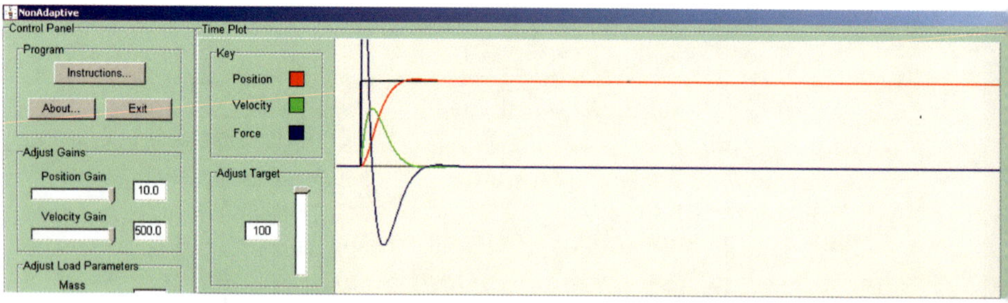

Figure 5-5. Effect of raising mass by 20 units (from the initial value of 0.2 units).

The main effect of this 100-fold increase in the mass of the load is not on the motion of the mass (red trace), but on the pattern of force applied to the mass by the hierarchical control system (blue trace). This is the dynamical analog of the effect we saw in Chapter 3 on static values of variables—in that case, the main effect of changing a gain factor of a box was to alter the static value of the variable at the input to the box rather than its output. Here, drastically increasing the mass of the load has very little effect on the motion shown by the load after a change in the reference signal: the primary effect is on the way the control system dynamically applies forces to the mass.

With this much greater mass, we can now see the initial spike of applied force very easily—it goes off the top of the plot. And the subsequent braking effect is clearly shown as the blue line representing force drops immediately far below zero, and then slowly becomes less negative, reaching zero just as the mass reaches the reference position. The mass now overshoots the reference position by a small amount, but only a small amount and only for a moment.

It certainly looks as if the control system has altered its own characteristics to compensate for the increase in mass, but it has not. Its only adjustable characteristics here are the two gain factors, and they remained as they were before increasing the mass. The control system has not adapted its properties to the changed load at all. What we are really seeing are reflected effects of the position and velocity feedbacks.

You can verify that changes in the spring constant and the damping constant likewise have only very small effects on the behavior of the mass, and have much larger effects on the pattern of applied force (though less dramatic than the effects of changing the mass).

You can click UPDATE BODE PLOTS (see Figure 5-3) to initiate an automatic experiment of the kind done by McReur and Jex, in which the load parameters are changed and then a frequency-response curve is obtained, with the measured error signal being the input and the force applied to the load being the measured output. First only the spring constant is set to maximum with the other two parameters at zero; then only the damping constant is set to maximum, and finally only the mass. The sliders that vary mass, spring, and damping are ignored during this process. Changing the Position and Velocity gain sliders does alter the plots. For each condition, the reference signal is varied in a sine-wave starting at a frequency of 0.1 radian per second, for ten full cycles (which is sufficient for the output sine-wave amplitude to become steady). Then the frequency is increased a little and the new output amplitude is found, and so on until the input frequency is about 12 radians per second (1 cycle per second equals 6.28 radians per second). This covers the same frequency range as in the old experiments. On a fast computer this all happens very quickly in vastly speeded-up simulated time. During this process, the phase relationship between the error signal's sine wave and the output sine wave is also measured. The gain of the system at any frequency (plotted on a logarithmic scale) is the ratio of output amplitude over input amplitude, in decibels which are also a logarithmic measure, and the phase angle is the number of degrees of phase shift between output and input. (If you don't know what phase shift is, don't worry about it.) We see one plot for each condition of the load in Figure 5-6.

These plots closely resemble the experimental results obtain by McRuer and Jex, using human subjects and slightly different variables. Ever since 1967 it has been assumed that they were measuring actual changes in the properties of human control systems. We can see now that at least a large part of the observed changes need not be attributed to adaptation of the human systems, because if the human system contains a two-level hierarchical control system of the kind we see here, no adaptation is necessary to produce these results.

I emphasize that the point of this demonstration is not to do away with adaptation; it is to show that negative feedback control systems can be designed very simply so they operate over a wide range of environmental properties without requiring any adaptation at all.

Adaptation is needed only when the changes in environmental properties become greater than what we have seen here:

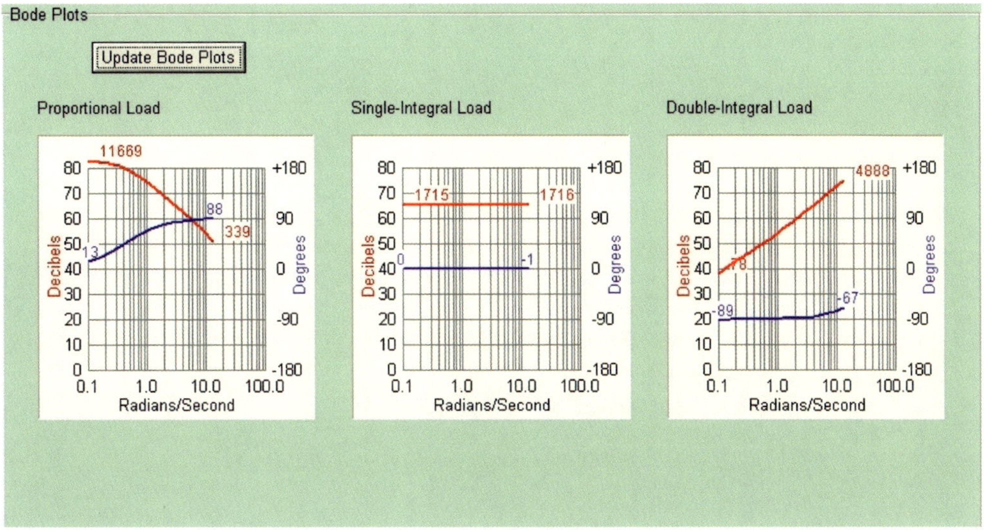

Figure 5-6. Bode plots for spring alone (left), damping alone (middle) and mass alone (right). Input is the reference signal, output is the force, measured in the frequency domain. The apparent characteristics of the system change from a falling frequency response to a level frequency response to a rising frequency response—but there are no changes in the control system's properties at all. Plots are shown for maximum position and velocity gain.

References

McRuer, D.T. & Jex, H. R. (1967). A review of Quasi-Linear Pilot Models. IEEE Transactions on Human Factors, 8, 231-249.

See also Chapter 14 in Jagacinsky, R. J. & Flach, J. M. (2003). Control theory for Humans (Mahwah, NJ, Erlbaum Associates).

Chapter 6: Multidimensional Control

We turn now from control systems stacked into a hierarchy to control systems all at one level, operating in parallel but controlling in different dimensions of the local world. The basic approach to control in many dimensions that we will take here is elementary, but it is a start toward understanding large organizations made of control systems.

A *dimension* is, mathematically, simply a way that something can be changed without altering it in any of the other ways it can change (my apologies to mathematicians). An object in space, for example, can move left and right relative to an observer, and up and down, and closer and farther away. Those are the familiar three dimensions of space. But the same object can also turn around any of three axes at right angles to each other, and it can have its shape, its color, its surface texture, its temperature, and its price changed, each way of changing being at first glance independent of the other ways. If we want to control all these aspects[24] of the object, the problem is solved rather simply, it would seem, by assigning one control system to sense and act on the object in each independent dimension.

24 from aspectus, past participle of aspicere, to look at. An aspect of something is a way of looking at it, not a property of the something.

There is always, however, a "however." The snag is in the word independent. Even when the aspects of an object seem to change independently, physics and other kinds of natural and invented law sneak into the process, so that changing in one way creates side-effects that tend to change the other aspects. This is not too hard to understand. Consider color and price, two attributes[25] of an object that we might feel free to change independently of each other. If the object is a sirloin steak, however, and its color changes from red to dull green, its selling price will be severely disturbed.

This is the general problem of multidimensional control. What seem to be independent dimensions of change are actually coupled to each other, sometimes only loosely but sometimes very strongly, through pathways that are not obvious. This means when a control system is used to alter the object in one dimension, doing so creates disturbances of one or more of the other dimensions. The opposing actions taken by the other control systems can in turn disturb the first control system, so an unintended feedback loop can be created. When systems become coupled together through these accidental interactions, the most common result is instability, because two symmetrically interacting negative feedback systems form a positive feedback loop—an error-*increasing* loop.

There are ways around this problem. The most obvious way is to design the systems so that the coordinates of the variables to be controlled are orthogonal to each other. That just means that each variable can be controlled at any reference level without disturbing the others. If this can't be achieved perfectly, it may be possible to minimize the interactions enough. Enough means that each control system can counteract the disturbances caused by changes in the other dimensions without creating counter-disturbances so large as to make the whole multidimensional system unstable.

25 from ad + tribuere, to bestow. An attribute of something is created by the act of attributing, or bestowing, a property to or on it; therefore an aspect, a perception, not a property.

Another way that works when the physical system itself can't be made fully orthogonal is to create orthogonality by the way the individual control systems sense and affect the multidimensional object. If an action used to control one variable disturbs another control system, the action can be given a second effect that tends to cancel the disturbance of the other system's controlled variable. This "feed-forward" kind of effect can be, in practical terms, only approximate, but it can make the difference between stability and instability.

It's also possible to achieve better orthogonality by changing the definition of what the control systems are sensing and controlling. This is done by letting each control system sense and act in more than one dimension.

The dynamics of multidimensional control can be complex, but that subject is beyond the scope of this book (and its author). We know from Chapter 5 that a rather simple two-level system can create stable control of a load with a wide range of dynamical characteristics, so perhaps with the right approach the subject of dynamics will not prove too difficult when someone is ready to tackle it from the present point of view. For now, we will not consider dynamics, except to make sure that each control system by itself would be dynamically stable.

TWO SYSTEMS, TWO VARIABLES

Let's start with a very simple control problem. Figure 6-1 shows two perceptual input functions. Each system senses the states of both variables, and each system's actions affect both variables. The problem: to construct an input function and an output function such that each system can independently sense and control some aspect of the environment. How could these two variables be combined in two different ways that would generate two independent perceptual signals? In the simplest case, this boils down to the question of how the signals might be weighted before being added together to produce a perceptual signal.

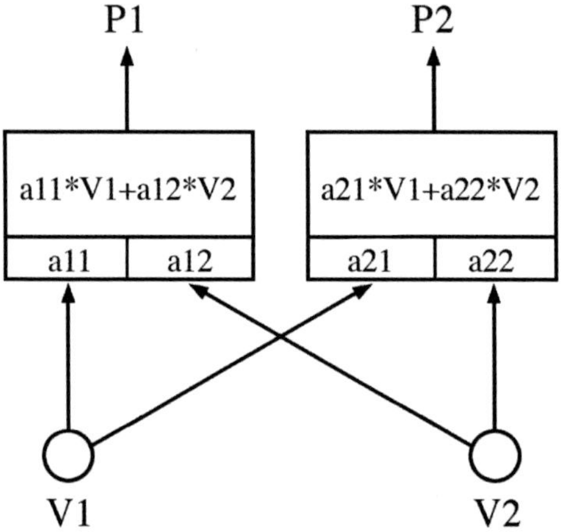

Figure 6-1. Two perceptual input functions creating perceptual signals (P1 and P2) representing different aspects of a two-variable environment (V1 and V2). The weighting factors a11 ... a22 determine the aspects that are represented as perceptual signals.

The simplest weightings would result from setting a12 and a21 to zero. Then P1 would represent V1 alone, and P2 would represent V2 alone. But suppose that for some reason this could not be done—suppose that the input function on the left formed with random settings for its input weights, and the same for the input function on the right. In that case the cross-connection weightings would almost certainly not be zero, and precise orthogonality of these two input functions would be unlikely. The first question is what will happen if we just take the two input functions as they are, random weights and all, and try to construct two control systems using them. What we're doing here is seeing what sorts of problems might arise in the evolution or learning of new control systems, when there is no external guide making sure it all comes out right.

To create a control system, we need a comparator that receives a reference signal from somewhere and a perceptual signal from one of the input functions. The comparator produces an error signal that enters an output function, and the output function generates an output action that affects both of the environmental variables through more weighting factors. We assume that this basic architecture exists or strongly tends to form during development of this system, so the only unknowns remaining to be determined are the magnitudes of the various weightings. Figure 6-2 shows the design for the two control systems we are considering:

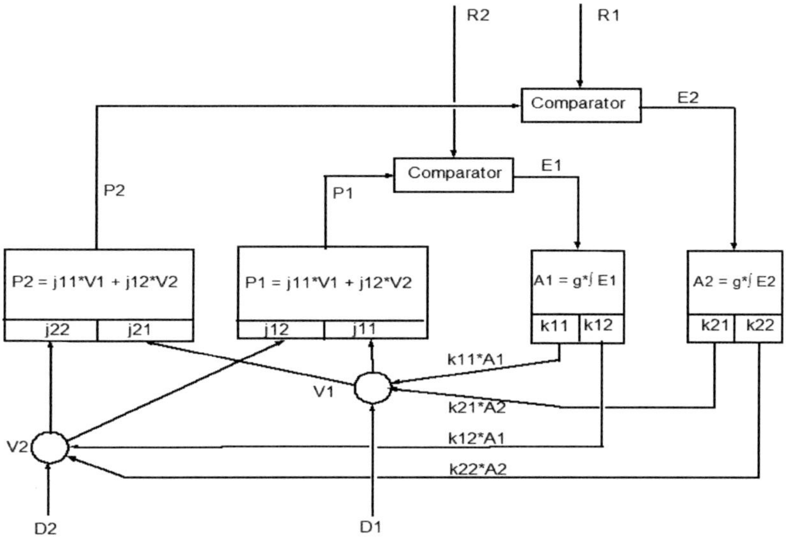

*Figure 6-2. Two control systems. Each one senses a weighted sum of two environmental variables V1 and V2 and acts on both. The output effects add where they act on V1 and V2. Two disturbances, D1 and D2, also affect V1 and V2 additively. The output functions generate an output action Ax (where x is 1 or 2) that is the integral of the error signal. The action is multiplied by an output weight to determine its contribution to the state of each environmental variable. The state of the environmental variable is the sum of the actions and the disturbance affecting it: Vx = Dx + k1x*A1 + K2x*A2.*

The control systems are given integrating output functions; this is a simple way to make sure they are dynamically stable, since all other functions are simple proportionalities.

Assuming all the weightings and the output gain to be positive numbers, we can see that negative feedback will exist if the comparator subtracts the perceptual signal from the reference signal (rather than the other way around). That puts a single negative sign into the loop. If some input weights are negative, then the corresponding output weights will also have to be negative to preserve negative feedback.

The easy way to achieve this is just to make k11 = j11, k12 = j12, k21 = j21, and k22 = j22. This would hold true no matter how many systems there were, given that the number of environmental variables equals the number of systems.[26]

Of course this does not tell us how, in a living system, the output weights could possibly be made to correspond correctly to the input weights, which exist physically in a different place as various synaptic strengths in different neurons. All we have found is that if this condition, or something like it, could somehow be achieved, all of the feedback loops would come out to have negative loop gain, assuring negative feedback, one of the requirements for stable control. After we have looked at a possible mechanism for adaptation—I call it Reorganization—we will see whether this process can achieve something close to the required relationship without divine intervention. That will be the subject of the next two chapters.

For now, we will simply use the rule that the output weights are copies of the input weights related to the same environmental variable and not worry about how that condition is achieved. Also, to avoid other complications I have adjusted the triples of input weights so the sum of the squares of their magnitudes is 1. When we do all this in a different way in the next Chapter 7, that won't be necessary. Normalization, as this process is called, keeps the control sensitivity within reasonable bounds in the absence of any adaptive abilities (which will be seen in Chapter 8).

EXPLORING MULTIDIMENSIONAL CONTROL

- It's time to start the program called Demo 6-1.

This is another Live Block Diagram, this time showing three control systems with the same general organization as in Figure 6-2, but with one more system and environmental variable. There are sliders for adjusting the three reference signals, and sliders for adjusting three

26 I'm informed that a more conventional way of numbering the indices would show that the output matrix is the transpose of the input matrix. See Kennaway's Appendix.

disturbances. The output action, labeled ACTION, is shown as a number for each system, and the values of the three environmental variables and the three perceptual and error signals are also shown. The program is initialized with random weights between -1 and 1 for the three weights in each of the input functions (normalized, as just mentioned), and the weights in the output function of each system are automatically set to be copies of the input weights.

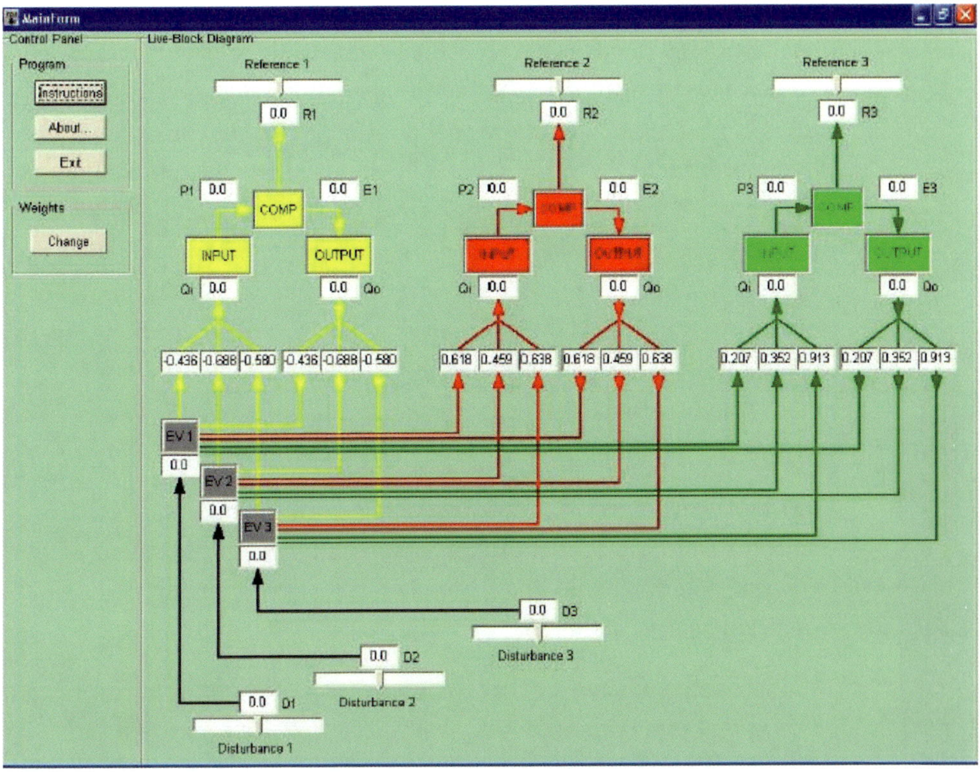

Figure 6-3. Three simple control systems sharing control of three environmental variables (EV). Reference signals and disturbances are adjustable. New sets of random weights (normalized) can be produced by clicking the CHANGE button in the control panel. For each output connection from a system (yellow, red, or green) to an environmental variable, there is an input connection from the same environmental variable. The output weights and corresponding input weights are equal. Qi is a virtual controlled quantity, not actually present as a single external entity. Speed of control depends on degree of independence of input weightings.

You can now verify that changing each reference signal will cause the corresponding perceptual signal to change without materially changing the other two perceptual signals. These three systems manage to control their own perceived aspects of the environment independently of the other two, despite interactions among a set of environmental variables sensed by all three systems and affected by the actions of all three systems.

As you adjust the reference signal to make a particular perceptual signal change, *all three environmental variables* change. No one of them changes in exactly the same way the perceptual signal changes. Changing any disturbance ends up affecting all three environmental variables, as the other systems act to correct the errors they experience. One perceptual signal does not represent any one of the environmental variables. It represents all three variables. Since the randomly generated weights in the three input functions are highly unlikely to be the same across all three, each perceptual signal represents a different function of all three environmental variables. That is what is meant by saying the systems perceive different aspects of the variables. The weights bestow those attributes on the set of environmental variables.

It is possible, but unlikely, that these functions are orthogonal (independent). That is because the weights have been assigned randomly, and the probability of accidentally generating the combination of weights that results in exact orthogonality is essentially zero.

Another way you can see that the input functions are not creating orthogonal perceptions is to look at the output actions of all the systems. When you alter one reference signal, you can see the action of that system changing—but you can also see the other two actions changing. This is because the action that changes the perception you're controlling disturbs the perceptions in the other two systems, and they alter their actions as needed to keep their perceptions from changing. The actions are an indication that the corresponding perceptions are being disturbed.

As the caption said, when you click CHANGE button in the Weights panel, a new set of input weights is randomly generated and the output weights are set to equal the new input weights. You will find that with each new set, the degree of orthogonality of the perceptions changes. You can tell when orthogonality is low, because there is a lot more mutual disturbance among the systems; the actions needed to control the perceptions tend to be much larger and the error correction takes a lot longer. Some new weightings are so nonorthogonal that the action numbers become rather enormous—two or more systems are approaching a condition of direct conflict. The actions are limited to magnitudes of plus or minus 2000. The chances that a pure conflict will occur are almost zero—pure conflict would occur when two systems control in exactly the same direction, so at least one environmental variable (EV) would have to have two different values at the same time. But the organism made of these three systems is in trouble whenever one or more actions becomes very large.

Some implications of this model

Perhaps it is becoming apparent that these control systems are doing something rather strange. They live in a totally unstructured environment consisting only of variables which do not directly affect each other. It can be shown that this holds just as well in an unstructured environment with 500, or 5 million, control systems and environmental variables. So what are these aspects of the environment that these systems are controlling, so any one of them can be varied independently of the others just by varying one reference signal?

They are products of the way the control systems perceive their world.

The weights in the input functions of the control systems quite literally *create* the variables that this system perceives. Those variables do not exist in the environment; they are manufactured by whatever assigns the weights in the input functions. Yet they can be disturbed by real physical influences, and they can be controlled by variations in the physical actions of the control systems. Even though they have no physical existence in the absence of the control systems, they are perfectly real. They are shown as virtual input quantities, Qi, in the diagram, because they exist in perception, but not really in the environment.

A burning question that has vexed philosophers for a long time—because they have suspected that a situation something like what we see here exists—can be somewhat reframed in the light of these observations. The epistemological question can be put this way: if the environment were structured, is there any reason to suppose that the multidimensional perceptual systems in an organism would somehow possess input functions such that the perceptual signals reflect significant natural properties ("natural kinds") of the environment? Or is it possible that the organism will simply impose its own idiosyncratic organization as we see the control systems here doing? In short, is the world we perceive and control related in any understandable way to whatever actually exists on the other side of our sensory receptors?

A possible answer to these questions may exist in a process that I call *Reorganization*, which enables a relatively (but far from completely) unorganized system to gain control over its perceptions of the environment—by creating and changing its perceptual input functions, and by altering the connectivity of its output functions. We will take up that subject in the next chapter, before going on to multilayered and multidimensional control systems.

Chapter 7: E. Coli (Random Walk) Reorganization

When I first began working on a control-theoretic model with the late Robert K. Clark in the 1950s, it became clear to me that some gaps had been overlooked in the concept of learning that psychologists accepted. A basic question about living control systems is How do they get that way?—how do they learn to control things like tractors and computers and musical instruments and baseballs that have not been in existence long enough for evolution to have contributed anything helpful. Psychologists liked to study what they called "learning curves" that showed how an animal or a person gradually acquires a skill. The problem was and is that this process looks as if the organism knows what skill it is trying to learn before it learns it, and simply gets better and better at it before actually succeeding enough to be "rewarded" (or to survive and reproduce).

The phenomenon that seemed to define the problem for me was B. F. Skinner's demonstration of "shaping," in which he showed that by administering reinforcers—bits of food perceived through a theoretical filter—to pigeons, he could get them to execute alternating left and right turns while walking, so they followed a figure-eight pattern. In what universe could a pigeon have genetic instructions telling it that walking in a figure eight when hungry might produce food? For that matter, how could Skinner know (assuming he operated according to the same properties he attributed to his animal subjects) what final pattern of his own behavior would have the ultimate result of a pigeon walking in figure-eights? How did he know the pigeon would end up walking that way? Skinner was as baffled by this as I was, concluding that by some unknown sort of mechanism, behavior (his or the pigeon's) is controlled by its consequences. This usage of the term control was not, of course, the technical meaning in use here. Where did the consequence come from, if it didn't initially exist?

I saw it in a different way. To me, this was evidence of some very general mechanism inside the pigeon that allowed it to discover behaviors that would have effects that mattered to it. It was somehow finding out what action made Skinner produce food, and doing it. That was much too intellectual a way of describing the mechanism, but it put the responsibility for walking in figure eights where it belonged, in the pigeon rather than the behaviorist. All the behaviorist had to do was obey a rule saying that any move by the pigeon that brought its path closer to what he perceived as a figure eight was to be followed by delivering some food to the pigeon. The pigeon, however, had to discover what it was, among all the possible rules and laws of its universe and all the immediate consequences of its actions, that sometimes made the food appear. And it had to change itself again and again, until it was doing something that made the food appear almost every time. If it couldn't do that, shaping wouldn't work. Whatever mechanism in the pigeon was responsible for this convergence to doing the right thing must be enormously powerful, and at the same time inheritable and not dependent on logic or reason of any human kind. Otherwise, applying Skinner's capricious rule about what he wanted to see the pigeon doing before feeding it would just result in a skinny or seriously dead pigeon.

What was needed to account for this kind of learning was something basically simple, so it would work from the very beginning of a human or pigeon life, yet something that could discover relationships that were absolutely arbitrary, making no sense under any known organizing principle or law of nature.

The only solution I could see was a combination of a random change process and some process for selecting the result that was needed or wanted. Trial and error, repeated until the error got smaller. Others had already had similar ideas. W. Ross Ashby (1952) in particular actually built an example as his "Uniselector," which inspired my thinking. Donald T. Campbell (1960) called it "Random variation and selective retention."

That was the "aha." Instead of looking for something like "reinforcement" that would somehow cause behavioral organization to change in a specific way if the result was favorable, I was now looking for a mechanism that would cause organization to *keep changing as long as the result was unfavorable*. A favorable result then would let the process of change slow down or stop.

The basic idea is not one of searching through different behaviors that might work, but searching through different organizations of the control systems that produce behavior, at the level of changing the characteristics of these systems—the weighting of different inputs to the sensors, the gains and time constants of the output functions, the connections from one system to another or to the environment. This is a much more fundamental kind of change than simply choosing to do one already-learned behavior or a different one. It is changing *how* behavior is produced rather than *what* behavior is produced. Of course changing the *how* can change the *what* and vice versa.

It quickly became apparent that the process of change itself could be unsystematic and still achieve systematic results. An old observation is that the thing you need is always found, by some mysterious natural law, in the last place you look for it. If there is some condition in the organism that causes changes in behavioral organization to occur, and if removing this condition stops those changes, then of course the organization of behavior will remain as it was just before the condition was removed. It will look as if removing that condition is *reinforcing* because that removal, or the positive occurrence of whatever removed it, seemed to strengthen the tendency to produce that behavior from then on. What is really happening, according to reorganization theory, is that the tendency to *change* the behavior is *reduced*. The idea of reinforcement is the opposite of the process that I called *reorganization*. When I first grasped this concept, I used a trendy term for it, hoping to impress the cyberneticists of the 1950s: "negentropy." But all it meant was changing the organization of the behavioral systems or producing organization from scratch, both of which are included under the idea of reorganization. It was basically W. Ross Ashby's idea and Donald T. Campbell's succinct description, carried a little further.

While this was not quite a workable model, it did contain the seeds of one. It showed that if an error signal in some sort of very basic control system could cause outputs that randomly altered the organization of behavior, this reorganization would continue until a behavioral organization was produced that would bring the error signal to zero (if the organism didn't die first). Later, I called the set of all such control systems "The Reorganizing System." I also more or less abandoned this model because on second and third thought it just didn't seem efficient enough. How could the pigeon find out that it had to produce a sequence of left and right turns by trying all possible behavioral organizations at random? The probability of hitting on the right one in the first hundred years would seem to be about zero. It seemed that while the architecture of the reorganization model was fairly clear, a practical realization was not.

That was in the late 1950s. In 1980, Daniel Koshland published a little book called "Bacterial Chemotaxis" (1980) which provided the missing feature of the model of reorganization. It seems that the intestinal bacterium Escherichia coli, E. coli for short, can find its way up gradients of attractants and down gradients of repellents[27] even though it can't steer. It has flagellae which are spun by chemical nanomotors at their bases. When all these flagellae spin at the same

27 Note how "repellent" and "attractant" are defined by the way E. coli swims in gradients of different substances. This is an example of what Bateson called a "dormitive principle." The substances, of course, have no power to attract or repel bacteria. The bacterium actively seeks or avoids them.

time in the same direction, E. coli swims in a straight line. When some of the motors reverse, the flagellae spread out like a shaken string mop and E. coli's body tumbles in space. When the motors are synchronized again, E. coli takes off in a new direction that has been shown to be unrelated to the former direction. The tumbles verifiably produced new directions at random.

The trick is that when E. coli is swimming up a gradient of some good chemical in its surroundings, or down a gradient of a bad one, tumbling is delayed, while if the direction is toward the bad or away from the good, the time to the next tumble is shortened. The net result is that E. coli spends a lot more time swimming in the right (for it) direction than the wrong one. On average, it (or rather a simulation based on this principle) makes its way toward the source of the good substance or away from the bad one over half as fast as it could do if it could just turn and swim the right way.

Suddenly random reorganization didn't look as inefficient as I thought it would have to be. The key idea is that what is changed by a reorganization is not a position, not the value of some behavioral parameter, but the *rate of change* of position or value. And what was sensed was not the degree of goodness or badness of the result, but the first derivative of goodness or badness—whether the result was getting better or worse. When I tried that out in a model, it worked beautifully. Better yet, it worked *fast*, far faster than pure random trial and error, even though the changes were still truly random.

In E. coli what is reorganized is its direction of swimming, a measure of behavior, which is not the sort of reorganization meant here. But the bacterium's direction of swimming through space can be treated as analogous to the direction of altering a set of control system parameters in some abstract multidimensional space, and a tumble can be treated as an alteration in the proportions by which the parameters are being changed, so the direction of metaphorical swimming in this multidimensional space changes.[28] There will always be some location in this space, some specific set of values of the parameters, that will produce the best results possible in terms of the effect of behavior on something important to the organism. If there is a way to measure how far from that point in multidimensional space the system is, and whether the distance is increasing or decreasing (whether the important effect is getting better or worse relative to some reference level), the E. coli method of locomotion can be used as a pattern for a method of reorganizing. While there are many possible realizations of this architecture, we will use the simplest one, a naïve linear adaptation of the E. coli method.

IMITATING E. COLI

Demo 7-1 Ecoli is about reorganization. It allows you to play the part of E. coli swimming in a two dimensional space up a gradient of something to its source in a little circle on the computer screen. Ready?

- Start Demo 7-1. The boxed text in the control panel describes your task.
- Click RUN and try to steer the moving red dot to the circle.
- To change the dot's direction, click TUMBLE.

28 During a recent trip to a museum of natural history, I encountered a display of a series of dinosaur skeletons showing how the size changed through time from small to large while the skeletal details, like ribs, vertebrae, and limbs retained the same general form. It seemed pretty clear that I was looking at E. coli reorganization: the body shape was "swimming" in a direction that made it larger, and apparently this was also making survival more likely because the swimming kept on in that direction for millions of years without any "tumbles." Reorganization in a single organism, of course, doesn't take that long. But that may be the time scale appropriate to mitochondrial DNA, which is preserved across generations.

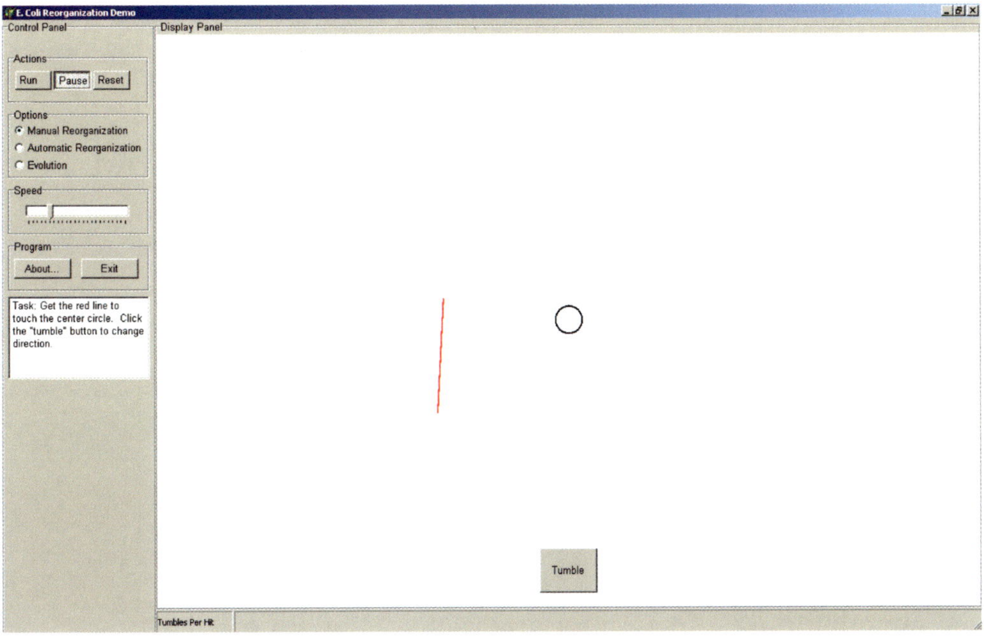

Figure 7-1. E. Coli Reorganization. Clicking the TUMBLE *button causes a random change of direction of the moving dot. The task: steer the dot to the circle in the center.*

When you start the demo, you see the circle at the center and a dot moving in some direction elsewhere on the screen, leaving a trail (if it goes off one edge it wraps around to the opposite edge). The object is to steer the moving dot into the circle. Your only means of steering is to click the TUMBLE button. After one tumble, the space bar on the keyboard will also work and is a little easier to use. Tapping the space bar causes the direction of motion of the dot to be changed at random: an E. coli tumble. While you can have runs of bad luck, most of the time you can get to the circle in under half a minute, using 15 or so tumbles on average. You are getting a completely systematic result by using a behavioral change that is completely random—except for choosing the time when it occurs. At the bottom of the screen the cumulative average ratio of tumbles to hits on the target is displayed.

You can click the **Automatic Reorganization** radio button to start an E. coli model running, with time speeded up. This model lets a little time go by after a tumble before judging whether the distance to the goal is increasing, which simulates the reaction-time delay you no doubt experienced. It will behave much as you did, though it may be a little more efficient (fewer tumbles per hit, fewer moves toward the target followed by an inadvertent tumble you didn't mean to cause).

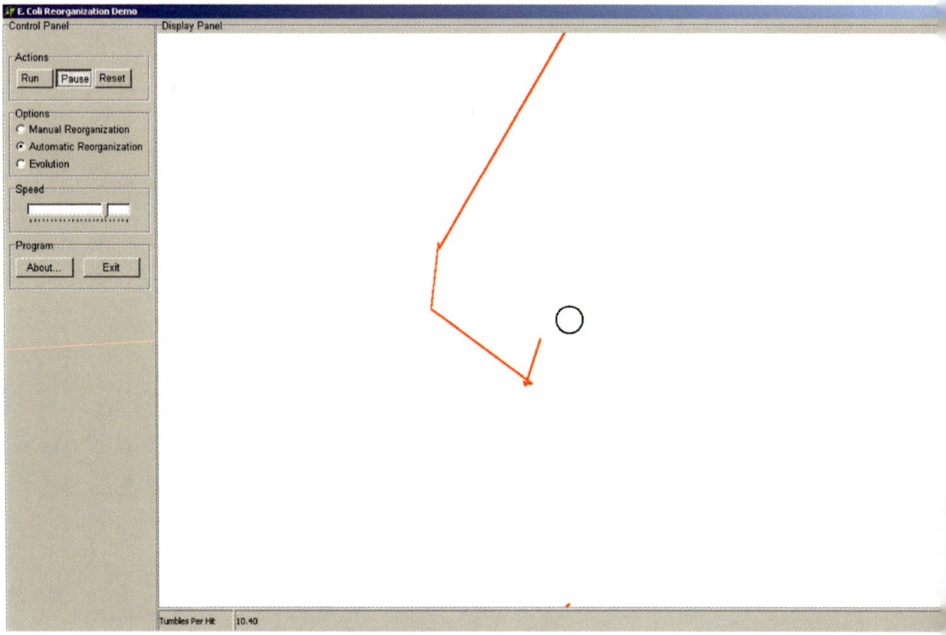

Figure 7-2. Automatic reorganization. Now the moving spot is under the control of a simulated E. Coli reorganizing system.

If tapping the space bar simply causes the dot to jump to some other random position on the screen (the third setting of the radio buttons does this automatically and very fast), the probability of landing somewhere in the circle is the ratio of the number of pixels in the circle to the total number on the screen. On my computer, the effective part of the screen in use has 598 x 598 pixel resolution, and the circle contains 314 pixels, so for each press on the space bar, the probability of a hit is 0.00088, and the probability of a miss is 0.99912. This tells you that there is a 50-50 chance that the dot will not hit the circle in 789 jumps. To get a 90% probability of success you would need, on the average, 2620 tries (10 per cent of the time you wouldn't get there in 2620 random jumps). If this were happening in three dimensions, the number of tries for 90% probability would rise to about 130,000; in four dimensions, to 7 million. As the autoreorganization mode shows, the average number of presses per success for E. coli reorganization is about 14 (compare with 789). So E. coli reorganization works, in this situation, about 50 times as fast as simple random mutations of parameters would work—in just two dimensions. It's actually much more efficient than that if you consider the 90 percent probability level.

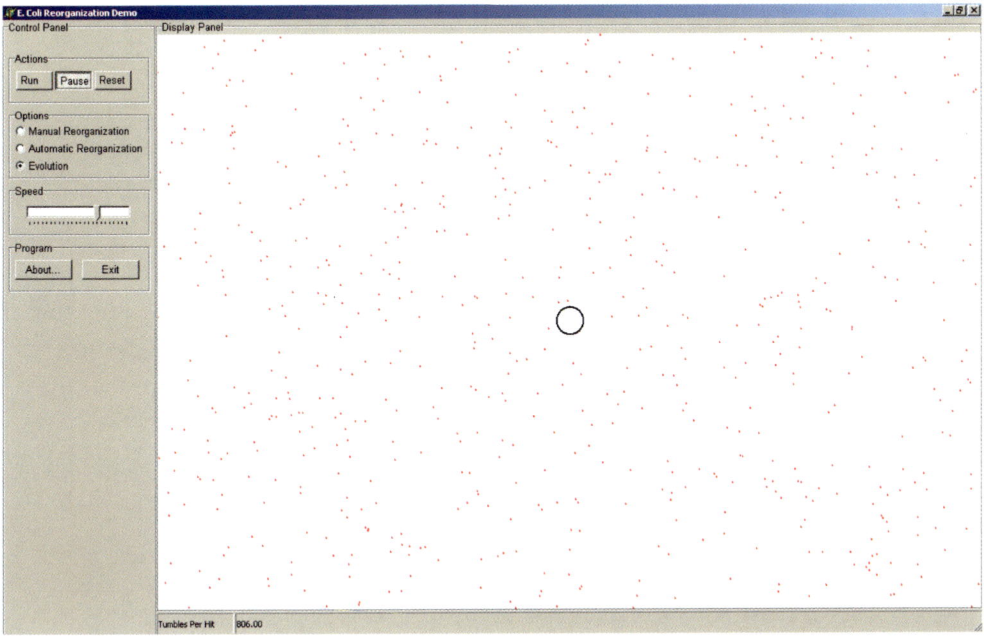

Figure 7-3. Mutation model: now each tumble simply causes the dot to move to another randomly-selected position. The probability of landing in the central circle in a given number of random changes (after extended runs) is about 1/50th of what it was with E. coli reorganization.

Each dimension represents one parameter of behavior that can be changed independently of the others. How many variable parameters are involved in a behavior pattern like walking in figure-eights?

When the number of parameters or dimensions becomes very large, even the E. coli method is defeated. There seems to be some number that is the maximum practical limit for this method. But that works out quite well for the PCT model, in which control is carried out by many independent one-dimensional control modules at many levels of organization. If each level reorganizes independently of the others, that reduces the number of dimensions being reorganized at the same time. If there are natural groupings of systems at each level, the problem is again made easier. And if this capacity to reorganize is distributed everywhere and works locally, we may have a sure bet.

Let us not worry about that. This is the model we have now; we can live with it until the next one comes along.

A demonstration of 3-D reorganization.

Let's try our model of reorganization with the multiple control systems of Chapter 6.

- Start Demo 7-2 now.

This is quite an elaborate demonstration, good for occupying a few long winter evenings. We will explore only some of its possibilities.

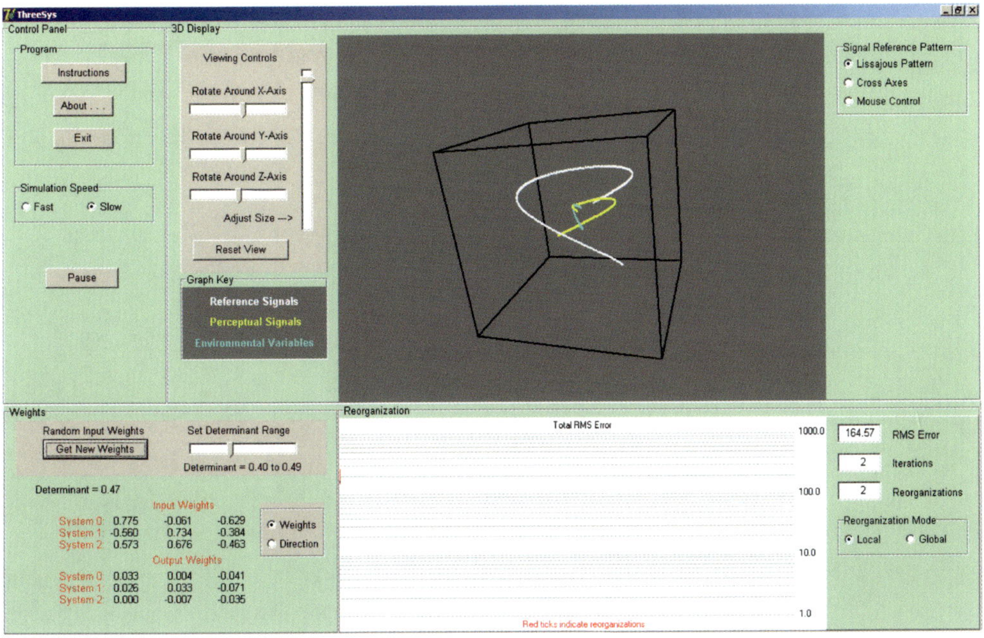

Figure 7-4. Moving white dot (leaving a curved trace) shows the reference signal changing in three control systems like those in Chapter 6. Each system's variables are plotted in one of the three dimensions of the display. The yellow trace shows the three perceptual signals, and the aqua trace the behavior of the three environmental variables in external (laboratory) space.

The fields and controls for this demonstration are grouped in sections on the screen: Control Panel, 3D Display, Signal Reference Pattern, Weights, and Reorganization.

We are using a three-dimensional control model, so our E. coli-type reorganizing system will be looking for error in all three control systems. It doesn't care what the error is about; error itself is bad. For simplicity we'll say this system is made of three independent reorganizers that perceive the error in each system (Local reorganization). That's a very crude criterion; in a real organism, the criterion will be the effect behavior is having on some variable (like circulating glucose) that is important to survival. But we are still at a crude stage of developing this model, so we'll use this simple criterion.

A three-dimensional image on the screen shows the values of three key behavioral variables: the reference signals, perceptual signals, and environmental variables. We can plot each variable as a point in three dimensions. If the point is moving parallel to the x axis of the figure, for example, that variable is changing in the control system that controls in the x direction while it is constant in the other two control systems.

The three reference signal values are represented by the x, y, and z positions of the white dot. A second dot (yellow) shows where the three perceptual signal values are in this space. The three environmental variables have values shown by the position of the aqua dot.

The reference signals will be changing in one of three ways. The default pattern has two of the reference signals varying as the sine and cosine of an angle that keeps changing through 360 degrees, while the third varies as the sine of twice the angle. This makes a nice curved, though meaningless, Lissajous pattern inside the cube on the screen. The second choice of pattern causes the reference signal for each system in turn to vary positive and negative, returning to zero, so the pattern is a three-D cross showing the three axes of motion. The third brings up three sliders which you can use to vary each reference signal arbitrarily, by itself. Choose the Lissajous, Cross Axis or Mouse Control pattern in the Signal Reference Pattern section of the screen. The following experiment assumes the default condition, the Lissajous pattern. We still start as in Chapter 6 with a random set of input weights for each control system, but now the output weights begin at zero instead of being set to equal the input weights. Each output function has three weights, which determine the effect of the output quantity (amount and sign) on each of the three environmental input variables. These output weights will be what the reorganizing process changes.

On the Control Panel, use the **Fast** and **Slow** radio buttons to adjust the speed of the display.

- Click **Fast** to speed up the display by a factor of about 50 (Figure 7-5).

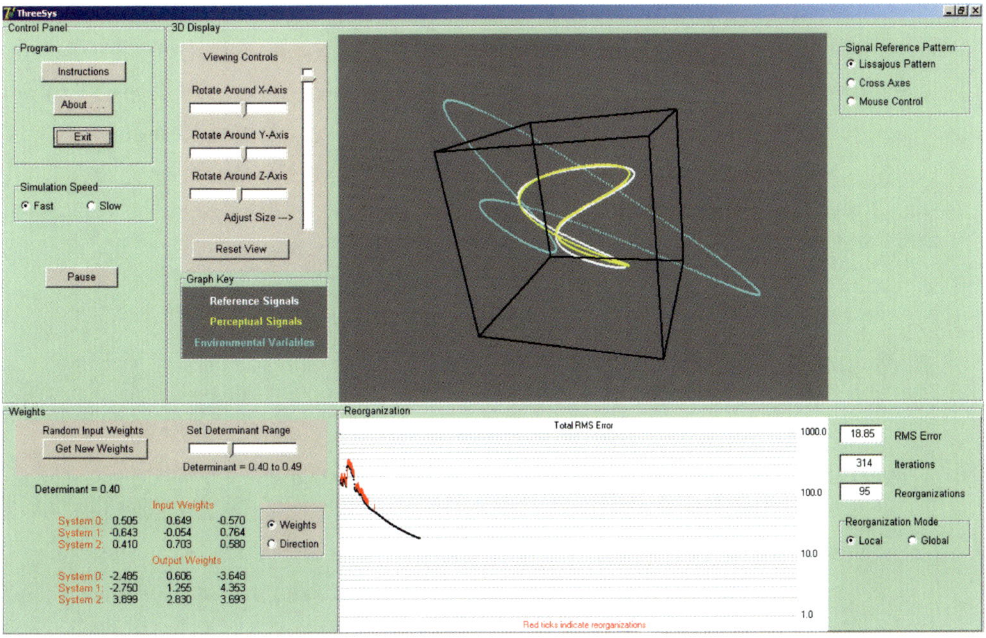

Figure 7-5. Fast mode. After about 300 iterations, the RMS error has become considerably smaller, with the yellow perceptual signals tracking the white reference signals much more closely. Note that the pattern made by the environmental variables (blue) is rotated relative to the perceptual and reference signal patterns.

Three sliders in the 3D display section are now active; use them to change your point of view in perspective so you can figure out what you're looking at.

Use the long slider next to the gray 3D model display area to change the viewing distance so you can see larger volumes.

The changes you saw at first are now happening at such an extremely rapid rate that a complete cycle of tracing out the patterns occurs something like 20 times per second. As a result what you see is a sort of writhing loop, or three writhing loops, as the weights change and the control improves. If you want to see what is really happening, click **Slow** to see the patterns slowly being traced out as the three reference signals change and the three perceptual signals more or less follow them (at first, very much less). Since the system is given a chance to reorganize only once per complete tracing of a figure, and the error plot is shown only after every complete tracing, there will not be many reorganizations. In the Fast mode, you will see that the yellow trace representing the three perceptual signals, after some rapid reorganizations and (sometimes) some fairly wild behavior, gradually begins to match the white trace representing the three reference signals.

The reorganization starts when the program starts. Each weight in each output function begins to change very slowly, in different directions (positive or negative) and with different speeds. Use the radio buttons in the Weights section to set the numerical display to represent either the input weightings or the direction in abstract space in which the output weights change on every iteration. The speed of changing decreases as the overall error decreases. The overall picture is that of E. coli, swimming in a three-dimensional space in a straight line. The space we're talking about now, however, relates to the weights, not the three dimensions we're looking at on the screen.

At the far right of the Reorganization section, the RMS value of the errors in all three control systems over one complete cycle of the reference signals is shown, and a plot in the center shows the total error on a logarithmic scale. As long as each control system's own error continues to decrease, the changes in its own output weights continue in the same direction, the same relative proportions (select Direction in the Weights section to see the velocities of change for each output weight). Eventually, the error in one system will reach a minimum and start to increase again, either because the optimal parameter value has been passed or because instability is beginning to appear. As soon as the error starts to increase by a detectable amount, the reorganizing system tumbles. A red mark on the error plot shows when that occurs. Since only the sum of errors is plotted, there's no indication of which system or systems tumbled. A tumble is a change in the direction and speed of change of each output weight.

In the Weights section, a table of nine numbers specifies speeds of change of the weights (this is the table displayed by the **Direction** radio button); tumbling sets the three of them corresponding to the system in question to values picked randomly between 1.0 and -1.0. This determines the new pattern of changes of those three output weights.[29] The system's output function "swims in a new direction" in the three-dimensional space defined by all the output weights. If the error starts decreasing again, no tumbles occur. But if the error still increases, another tumble occurs right away. As many consecutive tumbles as required occur, until the error is decreasing again. Obviously there is a very strong bias toward moving in the direction of zero error despite the inability to steer.

As the error decreases, the speed of movement needs to decrease, too, so as not to overshoot the condition of zero error. This is accomplished by multiplying the changes in the output weights by the summed RMS values of all the error signals. The rate of change of all the weights thus slows as zero error is approached, and change stops when zero error is reached (it never gets to zero if the reference signals change continuously).

An overall gain constant is set to make reorganization as rapid as possible without overshooting.

Eventually the tracking becomes essentially perfect. But you will also notice that the aqua pattern, representing what the environmental variables are doing, while a quite regular pattern, is not aligned with the orientation of the reference and perceptual signal patterns, does not have the same proportions, and has a different size.
The composite control system perceives three variables precisely following their changing intended positions; we can see, however, that what is perceived is not an accurate picture of what is really going on in the environment. This picture is not seen by any of the three control systems, which each perceive and control only one of

29 There is also a mode of reorganization in which the total squared error in all three systems is monitored, and in which a tumble consists of randomly setting all nine directional numbers to new values between 1 and -1. Click the **Global** radio button to switch from **Local**. There isn't much obvious difference, though the local mode seems faster.

the dimensions of the display—but you, in the position of a higher level of organization, can see it.

In the Weights section, you can click the GET NEW WEIGHTS button to generate a new set of input weights and reset to the starting conditions, including setting all the output weights to zero. Also, to the right of that button, there is a slider for selecting the desired range of values of the determinant of the matrix of input weights. A determinant of zero would mean that there is no way for all three perceptual signals to be controlled independently. The input weights are normalized so the sum of their squares is 1 in each system—this means that if the determinant of the 3 x 3 matrix is 1.00, the three systems are controlling orthogonal dimensions, so they do not disturb each other at all. For determinants between 1 and 0, the systems disturb each other more as the determinant approaches zero.

As you will see, if you select random input weights with a determinant near zero, the systems will disturb each other a lot, the result being that the output actions conflict and therefore become very large (conflict has an effect like reducing the gain of the feedback function, which causes the error to increase, rather than decreasing the effect on the input quantity—see Chapter 3). The aqua pattern showing the variations of the environmental variables can become much larger than the viewing window, which is why there is a vertical slider next to that window for changing the distance of the figure from the observer. Also, when there is a lot of conflict, the reorganization becomes somewhat less efficient, and takes longer to bring the total RMS error close to zero. The efficiency, however, changes surprisingly little.

When the determinant is set to the highest range, 0.9—0.99, the pattern of the environmental variable changes gets as small as possible—but it still has a random orientation relative to the behavior of the reference signal and perceptual signal. So the perceptual pattern is still rotated in three dimensions relative to the reality, with a new random orientation each time a new set of input weights is created in the same range of determinant values. In fact, the input weights then almost form a rotation matrix, and the output weights reorganize to make another rotation matrix that just compensates for the input rotation. Now the perceptual signals give a picture of reality that is as close as possible to the truth, but still cannot represent the three angles of rotation. In short, the reality we experience is not very likely to be identical to the reality that exists. Of course we could add three more control systems to control the angular orientations—but then what dimensions would we still be ignoring, of which we can know nothing?

How does this relate to the original problem of the pigeon learning to walk in figure-eights? It solves it, in principle. We can now see how a series of random changes can be connected to some outcome that the pigeon wants (getting fed) so as to generate whatever organization of behavior is required by the current environment to produce that outcome. If the pigeon gradually alters its behavior in some "direction," it will get fed either more often or less often. If it gets fed less often, it "tumbles"—starts to change its behavioral parameters in some number of dimensions in a new direction. If it gets fed more often it keeps on changing in the original direction. Inevitably this will lead it to produce just the behavior that feeds it as fast as possible, or as fast as it needs (unless it starves to death first). It will be doing what B. F. Skinner wanted to see. What it sees itself doing might be, probably is, quite different from what B. F. Skinner's perceptions told him. I assume that there is no numeral 8 in a pigeon's perceptions.

For this to work, as Skinner found out, it's necessary to signal that the bird has done something closer to or farther from the required behavior. He did this by "rewarding" any move that was closer to a figure-eight pattern, and withholding what he interpreted as the reward if the move was away from that pattern. He emphasized the need for creativity in recognizing moves of the right kind, though he attributed no such ability to the pigeon. Skinner himself created the gradient in which this E. coli process took place. Without this more or less continuous (or at least repetitive) measure of progress toward a goal, there would have been no consistent relationship between what the pigeon was doing and the rate at which it got fed. Such a regular relationship is essential for the E. coli method of reorganization to work.

This suggests that there must be some underlying organization behind the process of reorganization, and that this organization must include indirect goals for behavior to accomplish. In line with W. Ross Ashby's concept of "essential variables," I proposed (in 1960) that there are certain internal variables, "intrinsic" variables, monitored by a set of reorganizing control systems, one control system per variable, and that for each such variable there is a built-in, inherited, reference level, possibly an adjustable one, or in some cases possibly a basic property of the detector such as a natural threshold (speculating about such details seems a little pointless at this early stage of understanding, and especially pointless since all that matters is the basic relationship between "intrinsic error" and random changes, however it is implemented).

It is this set of intrinsic variables and their reference levels that provide the measure of error required if E. coli reorganization is to work. These error measures allow the system to detect whether the result of a change is "better" or "worse." The intrinsic reference levels define the target of reorganization, like that little circle on the screen when you were imitating E. coli.

In our example of reorganization here, the intrinsic variables are the rates of change of the RMS values of error signals in the three behavioral control systems—"behavioral" to indicate that they are organized to use behavior to control perceptions of the local environment, and are not the reorganizing system. Each intrinsic variable in the present case is compared by the reorganizing system to a reference level of zero (any amount of any error signal is considered "bad"). This is a one-way control system: it acts only if the rate of change of this variable is positive—if the absolute value of the error signals is getting larger. In our model, the way it reacts is to produce a tumble, a random change in direction and speed of the way the behavioral control system's parameters are changing.

In the present example, the reorganizing system is controlling the quality of control exerted by a behavioral control system. When the absolute error in the three behavioral controllers is as small as possible, those controllers are controlling their respective perceptions as well as possible. Clearly, it would be possible to use other measures of intrinsic error, such as indirect effects of behaviorally controlling the three variables. It would be possible to reorganize in such a way as to control the fullness of the stomach, or crop, by varying the parameters of behavioral control systems that determine the direction, or curvature of direction, of walking.

What makes this reorganizing system so powerful is that it is not bound by logic, reason, past experience, or knowledge of natural law. It operates by pure trial and error. It doesn't have to know what the error signal means. It embodies what we mean by "trial and error" —try something, and if that doesn't get us quite what we want (error!), try something a little different. The key to success is the word "little." Make little changes in parameters, not big ones, if the error is little. There are limits to what this process can accomplish, but without it there could be no learning at all. Without reorganization, nothing new could happen except perhaps for the effects of cosmic rays.

One last thought before we move on. There are clearly two ways to conceive of random changes, or mutations, in organization. One is that the changes involve leaps like those of a grasshopper. The other is that they involve cautious movements like those of an inchworm. What the E. coli model seems to show is that the inchworm has the better strategy by a very large margin, two orders of magnitude better even for systems reorganizing in only two dimensions. And this may happen in the evolution of species, too. Successful evolutionary reorganizations are small and slow, not large and rapid. Large changes are invariably fatal; they produce monsters. I think evolution results from the capacity of organisms to reorganize at very low levels, in the genome. What I think, of course, may have little relation to the truth. But it's a possibility to be considered.

References

Ashby, W. R. (1952). Design for a Brain (New York: Wiley)

Campbell, D. T. (1960). Blind variation and selective retention in creative thought as in other knowledge processes. Psychological Review 67, 380-400.

Koshland, D. E. (1980). Bacterial Chemotaxis as a Model Behavioral System (New York: Raven Press)

Chapter 8: Multiple Control Systems at Multiple Levels

Now we will extend the model of reorganization beyond three dimensions, to show that it works as well in fourteen dimensions as in three. Then we will add a second layer of control systems to show how *coordinations* can be handled in this model, and a third to show how the two levels of coordinated control can be used to control a relationship of the arm to an independent object. The subject of reorganization will be left behind as we consider coordination, not because it doesn't apply but because it will take someone a little younger and possibly smarter than I to deal with all the problems involved. Reorganization will apply in the first demonstration that follows.

REORGANIZING IN FOURTEEN DIMENSIONS

Demo 8-1, Arm Control Reorganization, which I will now assume you have started, shows someone's disembodied right arm floating in space, as seen in Figure 8-1.

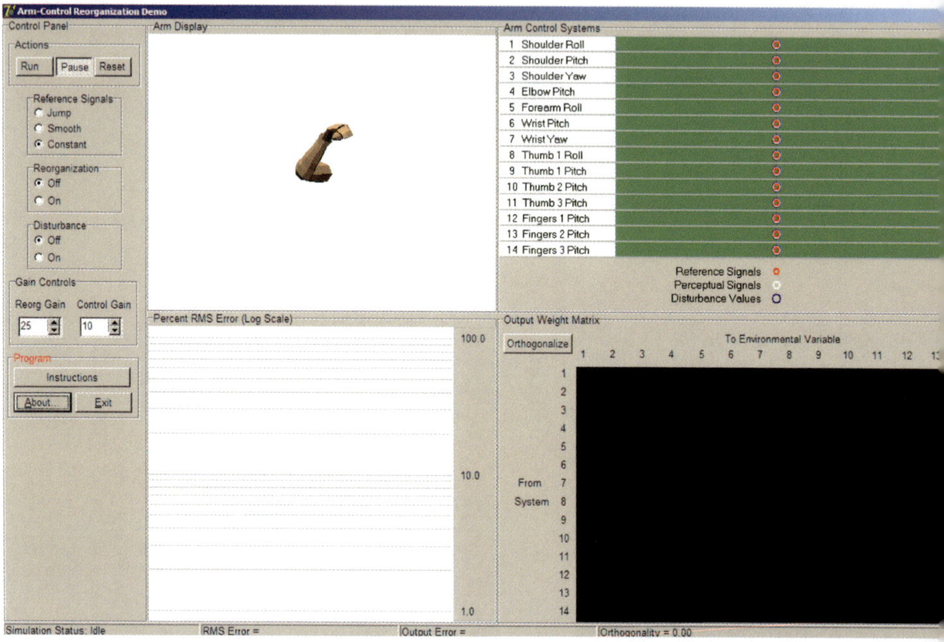

Figure 8-1. Arm with fourteen degrees of freedom.

- Click RUN (under Actions).

The arm will begin to move in a clumsy way. Now we are looking at *fourteen* control systems that control fourteen joint angles, all reorganizing independently at the same time, while their reference signals are varied through a smooth preset pattern and randomly varying disturbances are applied to each joint angle (disturbances were not used in the Three Systems demonstration in Chapter 7). The fingers move together as in a mitten. Let the demonstration run while you read on.

In each control system, the perceptual input function has provisions for fourteen weights, one for each of the 14 joint angles listed in the Arm Control Systems panel. However, only one of the weights (a different one in each system) is set to 1, while the other thirteen are set to zero.[30] This makes the perceptual signal in each control system represent a different joint angle. If we showed the fourteen sets of fourteen input weights as a matrix, only the diagonal entries, upper left to lower right, would be nonzero.

The reason for this somewhat over-elaborate way of setting up the input functions is to make it clear where, in the source code of the program, the input function might also be subject to reorganization. We won't do that here, but someone may want to try it later.

The fourteen output functions also have provisions for fourteen weights in each one, determining how much the output quantity of each control system affects each of the fourteen joint angles. This is clearly *not* how a real arm and nervous system would be organized, even in an infant, but it is a worst case. In fact, single joint angles in an arm are affected by several muscles, and also a single muscle can directly affect several joint angles, so the truth is somewhere between having one control system per joint angle and the extreme case we look at here. If reorganization will work in this extreme situation, it will surely work better and faster in the real one.

After 10 minutes or so the arm will be moving in a slow, irregular, but generally repeating pattern. The error plot, after some initial variations, is gradually trending downward as reorganization proceeds. The key to what is happening is shown in the 14 by 14 checkerboard (the Output Weight Matrix) at the lower right of your screen.

30 Here, a weight is a number that multiplies the effect of an input variable before it is added into the total to generate the perceptual signal. Giving an environmental variable a larger weight means it contributes more to the perceptual signal. A negative weight means the variable subtracts from the perceptual signal.

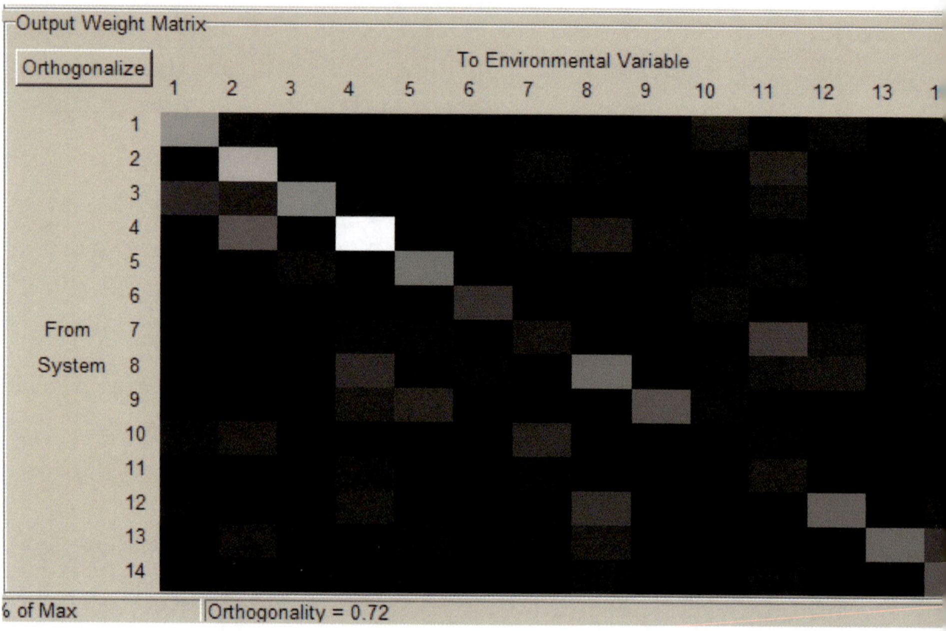

Figure 8-2. Brightness of rectangles shows weight connecting each output to each environmental variable.

In this checkerboard, the brightness of one small rectangle shows the relative value of one output weight in one control system's output function. A horizontal row of squares represents the fourteen output weights for one control system; scanning upward from one square to the number at the top of the checkerboard shows which environmental variable (joint angle) is affected through that weight. The output weights are all initially zero, so the checkerboard starts out black. When the largest output weight is less than some threshold amount, the brightness of each square is proportional to the weight represented by that square. After any weight has become large enough for its square to be white, the display is normalized to the largest weight. From then on, whichever weight is largest always appears white; the others then get darker as the largest weight gets larger. The numbered labels in the Arm Control Systems panel are the key to which control system outputs are indicated by the numbers down the left side and across the top of the checkerboard. For example, in Figure 8-2, the brightest white square shows that the fourth control system's output is having a relatively large effect on the fourth environmental variable, which is Elbow Pitch. The darker squares in the same row show there is little effect on the others.

For optimum performance, the output functions should become organized so that each control system's output affects only the one environmental variable it is perceiving. That would be the case if only the diagonal elements of the array, from upper left to lower right, were white and all the rest were black. This is the condition I refer to, somewhat inaccurately, as *orthogonality*. A number between 0 and 1 in the status bar at the bottom shows the degree of orthogonality of this matrix of output weights—it's simply the sum of squared weights along the diagonal divided by the sum of all the squared weights, with the square root then being calculated. After the demo has run for ten minutes the diagonal squares should be distinctly lighter than most of the others; after 15 or 20 minutes the contrast should be high. The orthogonality index, which starts at around 0.25, will be 0.70 or so and creeping upward toward the maximum possible 1.0. The reorganizing process is doing what it should do: reducing output effects that cause error and increasing mostly those that reduce error. To see what perfect orthogonality of the output matrix would produce without waiting,

- Click the ORTHOGONALIZE button. Use reset to start over after doing this.

In the Arm Control Systems panel in Figure 8-3 there are some green, red, and dark blue circles, one set for each of the 14 labels describing a motion about a joint (shown as concentric circles before the run begins). As Figure 8-3 shows, during the run the distance of the small green circles left or right of the centerline shows the perceived magnitude (relative to the center of its range) of the joint angle described by the label on the left. The slightly smaller red circle shows the reference signal's value for the associated control system. The dark blue circle, the largest one, shows the magnitude of a disturbance added directly to the joint angle. The reference signals are varying in smooth patterns. The disturbances are slowly varying in other smoothed but random patterns.

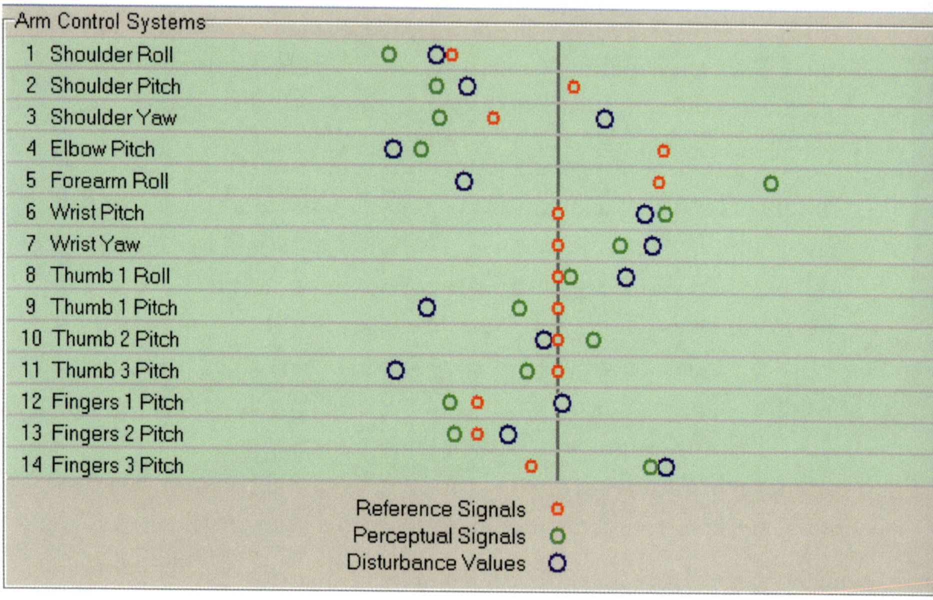

Figure 8-3. Performance of each of the fourteen control systems. They are named according to which variable they should be controlling.

At first, the disturbances have large effects on the joint angles, and control is very poor. The green circles representing sensed joint angles do not track the red reference signals, and the arm moves in uncontrolled ways. Each joint angle is affected not only by an independent disturbance, but by stray effects from all the other control systems. As time goes by, the green circles indicating joint angle follow the red reference signal circles more and more closely, while the effects of the dark blue disturbances on joint angle become smaller. By the time the orthogonality index has increased above 0.8, the arm is moving quite smoothly and the disturbances have almost no effect. Note that only the top five reference signals are changing.[31]

By clicking the **Constant** radio button in the Control Panel of the

31 If any readers are acquainted with Chinese physical arts, they will recognize the smooth movements of the arm (once they have become regular) as being part of a simple Tai Chi exercise. The resemblance was made just close enough to be suggestive. At a conference in Guangzhou, China, in 2006, an audience of Chinese scholars and students recognized the pattern as soon as it stabilized, and were amused.

main screen, you can freeze all the reference signals where they are. You can then left-click on a red reference signal circle, and holding the mouse button down, drag that reference signal alone to new values. Early in the process of reorganization, you will notice that changing one reference signal tends to alter many other green perceptual-signal circles—this is an effect of the off-diagonal weights which are still far from zero. This is easiest to study if you turn off the disturbances and the reorganization until you're finished experimenting. By the time the orthogonality index has reached 0.8 or more, changing one reference signal affects primarily the perceptual signal and joint angle for that system, the cross-system disturbing effects being almost gone. You can see that only one joint of the arm and hand moves, all the other joint angles being nearly unaffected. The lack of effect on other joint angles is partly due to the increasing orthogonality of the output matrix, and partly due to the increasing ability of each system to control its own sensed joint angle.

Off-diagonal elements of the checkerboard indicate an effect of an output coming from one control system on the joint angle being sensed by another one. All such effects are experienced as disturbances by the control systems controlling the affected joint angle, and the affected systems will produce added output to cancel the disturbance. Such outputs represent wasted energy; in a real arm they would represent several muscle forces being generated for the sole purpose of canceling the effect of another muscle force. As reorganization proceeds, this wasted energy becomes smaller and smaller, although the criterion for reorganization is error in the control system, not energy usage.[32] This criterion was arbitrarily

32 The subject of energy conservation, although irrelevant to this demonstration, may be important in understanding how the input function might be reorganized. As we saw in the previous 3-D demonstration, when the weights in the three input functions are set to approximate the condition of orthogonality, the perceptual signals become more and more independent. The output effects then disturb other control systems as little as possible. A condition of minimum internal conflict is reached when the three sets of input weights approach orthogonality. Therefore total energy usage could possibly work as a criterion for reorganizing the input functions to achieve independent dimensions of sensing.

selected for the purposes of this demonstration; others are certainly possible.

For reasons discussed earlier, we will now put the subject of reorganization aside, and explore the origin of the changing reference signals in Demo 8-1 that in this case produced the Tai Chi patterns. The patterns of reference-signal changes used in Demo 8-1 were created by simple equations selected to produce the needed results, with no pretense of modeling anything. Now we will turn to a more interesting way of generating the reference signal patterns needed for producing coordinated behavior of the arm segments.

COORDINATION

There are two ways to achieve coordination of joint angles to produce organized movements. One way is to calculate how all the joint angles must change to produce a desired motion of, let us say, the tip of a finger. This is called the *kinematic* approach and seems to be the mainstream idea as of this writing. It involves setting up a number of simultaneous equations and solving them (by operations involving inversions of matrices) for the patterns of joint angle changes that must be generated to produce a preselected pattern of movements, the trajectory of the fingertip. The machinery for doing these calculations, presumably some kind of computer in the brain, is part of the proposed model of coordination, although it is commonly left out of discussions about the complexity and plausibility of this model. When the physiological and physical properties of the arm, its mass and moments of inertia, as well as the neural computing functions needed to carry out the calculations are considered (the *dynamic* model), this calculate-and-execute model becomes markedly less plausible. This was discussed in Chapter 1, but I mention it again for the few who do not still have Chapter 1 vividly in memory.

The other way to achieve coordination uses the properties of negative feedback control systems to generate the required joint angles, and involves no precise computations or (precise or imprecise) matrix inversions. Demonstration 8-2, Coordination, shows how this works, and at the same time introduces a hierarchy of control systems with 14 two-layer systems in the lower level (like those in Chapter 5) and seven one-layer systems in the second level.

The aim of this demonstration is to show how coordinations can be created such that varying a single reference signal in a higher-order system can be used to generate a movement that requires variations in several joint angles at the same time. As in the demonstrations earlier in this chapter and in the previous chapter, there will be interactions among the second-order control systems, so that several different coordinations may be carried out at the same time while sharing the use of some of the first-order joint angle control systems.

What follows is only a very preliminary exploration of the possibilities of this kind of hierarchical control. Many questions will be left unanswered. I hope that the simple illustrations of coordination to be seen here will be sufficiently intriguing to entice others into further explorations of this kind.

In this demonstration, the arm still has the same 14 degrees of freedom as before. Each arm control system at the first level, however, is now organized as shown in Figure 5-2 (reproduced here),

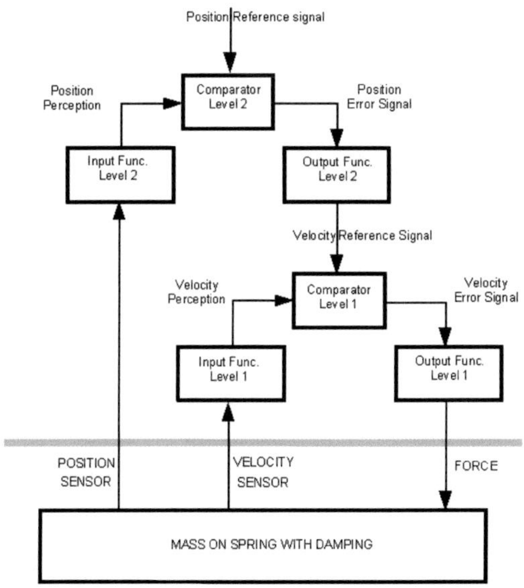

with velocity and position control (without damping or a spring but with mass) organized in a two-level hierarchy. In effect, each joint-angle control system now sees a load that consists of a free-floating mass (actually, since angular control is involved, a moment of inertia). This takes us a step toward a model with full physical dynamics, but leaves out dynamical interactions. From Chapter 5, we know that this kind of organization is robust with regard to changing dynamical characteristics of the load, so it's possible that this simple model would work adequately with a completely realistic model of the physics of this arm, including all the inertial effects and interactions. Time will tell.

The second level of joint control uses perceptual information about the arm's positions and orientations in objective (laboratory) space. Since nobody knows how the brain constructs perceptions of the visual world after the retinal image is formed, I simply use whatever information is available to me about the arm and create a perceptual signal in the model that corresponds to each relevant variable.

For example, the perceptual variables called hand pitch and hand yaw are generated by finding the x-y-z coordinates of two points on the hand. Hand pitch is obtained by subtracting one y-coordinate of a line segment along the axis of the hand from the other y-coordinate, at the other end. Hand yaw is obtained by comparing the z-axis coordinates of the ends of the same line segment. These measures break down at large values of the angles, but work over a reasonable range. They do not accurately measure the angles themselves (the measures are nonlinear functions of angle), but this doesn't matter, as you will see.

If you specify a reference pitch or yaw for the hand, the hand will be brought to a specific orientation which you can adjust as you wish.[33]

33 This simple way of generating the higher-level perceptions is one cause of my having to give up on adding reorganization at this level. The analytical geometry involved in preventing reversals of the sign of feedback for angles greater than 90 degrees—in 3-D space—defeated my remnants of knowledge of this subject, acquired 60 years ago. I decided that I needed to finish this book, a prudent decision at my age.

• Start Demo 8-2, Coordination. You'll see the image in Figure 8-4:

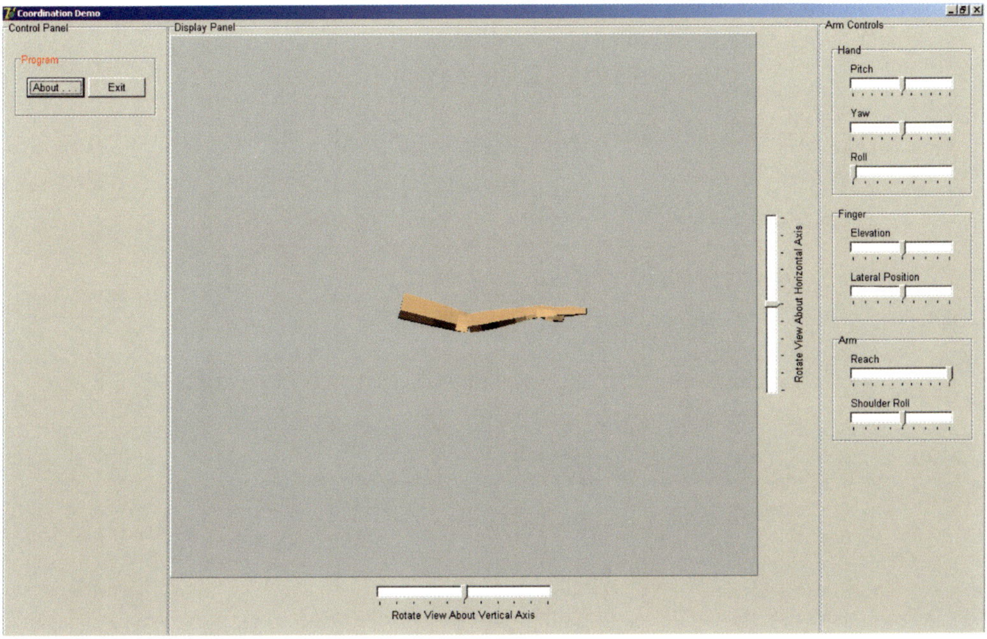

Figure 8-4. Arm with seven higher-order control systems. Reference signals for these control systems are set with the horizontal sliders.

The horizontal sliders adjust the reference signals for the seven control systems at the second level (the sliders near the arm display change the point of view). We will examine all of the second-order control systems to see how they work, starting with the REACH slider.

Reach: Control of distance from shoulder to fingertip. The controlled variable is the distance from the center of rotation of the shoulder joint to the center of the end of the mitten containing the four fingers being operated as a unit. This distance is computed from the x-y-z coordinates of the points at the shoulder and fingertips using the Pythagorean Theorem. The eye and brain, of course, do not use that theorem, but base their judgments of distance on binocular vision involving two 2-dimensional images on the retinas.

For programmers, the reach distance in Demo 8-2 is computed as

```
reach := sqrt(sqr(fingertip.x3d - shoulder.x3d) +
         sqr(fingertip.y3d - shoulder.y3d) +
         sqr(fingertip.z3d - shoulder.z3d));
```

That is the most complex part of the model of reach control. The rest of the model of that control system is contained in the next four program steps:

```
reachref := 100+reachslider/9;
reacherror := reachref - reach;
```

The reach error signal is translated into two first-order joint angle controller reference signals:

```
CSysLev1[2].r  := -(ReachError - HandElevError)/30;
CSysLev1[4].r  := -reacherror/30;
```

SysLev1[2] means the second control system at level 1, which is concerned with shoulder pitch angle. Its reference signal is indicated by the dot-r. Note that this reference signal is affected not only by the reach error signal, but by another error signal which comes from a hand elevation control system at the second level.

CSysLev1[4] refers to the level 1 control system for the elbow angle. The reference signal varies as the reach error at level 2 varies. Note that the scaling factor of 1/30 is the same as it is for the effect of the reach error signal on the shoulder pitch angle. This is not the optimum value—for a simple straight-line motion at the wrist, the elbow angle should vary twice as fast as the shoulder angle, as you can see by looking at your own arm. But this little scaling error of 100% does not bother this control system. In fact, the hand elevation control system takes care of the mis-scaling without even being told to do so. Let us proceed, logically, to look at that elevation control system and its counterpart in the horizontal plane.

Hand elevation: (vertical distance from shoulder), lateral position: (horizontal distance from shoulder). An alternative to these controlled variables would be the angle of a line from one eye to the fingertip in a vertical or horizontal plane. For purposes of this demonstration, the variables are computed this way:

```
HandElev := fingertip.y3d - shoulder.y3d;
HandElevError := HandElevRef - HandElev;
HandLateral := fingertip.z3d - shoulder.z3d;
HandLateralError := HandLateralRef - HandLateral;
```

Again, I simply propose that there are two perceptual signals, referred to as HandElev and HandLateral, which correspond to the actual difference in the y- or z-coordinates of the fingertip and the shoulder. I do not try to guess how such a signal is generated by the nervous system.

The error signals in these second-order control systems contribute to the magnitude of the reference signal for the first-order control systems, designated as CSysLev1[2].r and CSysLev1[1].r:

```
CSysLev1[2].r  := -(ReachError - HandElevError)/30; { repeated from above}
CSysLev1[1].r := -handlateralerror / 10.0;
```

The choices of signs in these equations are determined by two considerations: the definition of the perceptual signal in terms of a difference between two positions, and the need to preserve overall negative feedback. I confess that instead of working through all the details of the equations I simply changed the signs until control was achieved—my own reorganization process.

- Now you can try varying the reach distance using the slider.

You will see that the fingertip moves toward and away from the shoulder along a not-very-straight line. If you vary the HAND ELEVATION slider the position of the finger tip will move vertically while remaining at a constant distance from the shoulder because of the reach-distance controller. The HAND LATERAL POS slider moves the hand in the z direction, in and out of the screen. Use the view slider to get a top view that shows the lateral positioning.

Hand Orientation: Pitch, Yaw, and Roll. You will have noticed that during these manipulations, the hand stayed level as the arm moved.

This is because of two more control systems which sense and control the pitch and yaw angles of the hand—actually, the angle of a line between two points called *knucklebase* and *wristbase* located along the center of the back of the hand in the plane of the hand. The assumption is that the controller selects two points visually, representing as perceptual signals the pitch and yaw angles of the line between the points. Having no idea of how the brain could possibly do this but seeing that something very much like this must be accomplished, I simply say that the two perceptual signals exist. The angle measures are very approximate. These four program statements describe two second-order control systems.

```
handpitch := knucklebase.y3d - wristbase.y3d;
handpitcherror := handpitchref - handpitch;
handyaw := knucklebase.z3d - wristbase.z3d;
handyawerror := handyawref - handyaw;
```

The angle of roll around the hand's long axis is obtained differently

```
handroll := shldr + forer;
handrollerror := handrollref - handroll;
```

This is an approximate measure of the hand's roll angle, being simply the sum of the shoulder roll angle (shldr) and the forearm roll angle (forer), taking no account of the fact that there is a variable angle at the elbow joint getting in the way. As you will see, this makes no important difference in how you use the sliders to control the orientation of the hand. The control systems easily compensate for the nonlinearities.

The hand pitch and hand yaw control systems use two first-order control systems as follows:

```
sinr :=  sin(handroll);
cosr :=  cos(handroll);
CSysLev1[6]. r  := -0.02*handpitcherror*sinr - 0.2*handyawerror*cosr;
CSysLev1[7].r  := 0.2*handyawerror*sinr - 0.02*handpitcherror*cosr;
```

The 6th control system of the lowest order controls the pitch angle at the wrist joint, and the 7th control system controls the yaw angle at that joint. But you will see that the hand orientation systems at the second level each use *both* the pitch and yaw control systems at the wrist. By this means, the second-order control systems control the pitch and yaw of a line passing from wrist to fingertip, without regard to the roll angle of the hand.

If you set up the arm with some modest angle at the elbow and with the hand level, you will be able to see the effect of these control systems by operating the HAND ROLL slider. You can make the hand roll from palm-down to palm-up and see that the axis of rotation remains nearly constant even though the pitch and roll angles at the wrist are changing. Raising and lowering the hand, or changing its reach distance, will have little effect on the orientation of the roll axis if no limits are reached, and you can make the hand roll any way you like. You can also alter the hand pitch and hand yaw orientations and see that the roll axis changes accordingly. There are limits to the range of hand-arm configurations over which this coordination works well, but a little experimentation will show that you have similar problems producing the same effects with your own arm and hand. The model was given about the same limits on the joint angles that a real arm has.

My intention to keep all the control systems very simple was frustrated by the need to change the relative weighting of outputs from the second-order systems to the first-order reference signals as the hand roll angle changes. Sines and cosines are not the sorts of functions one expects a neuron to compute. On the other hand the exact forms of the functions used to change the weightings as the hand rolls are not critical; all that is really required is that the sign of the feedback effects remain negative. If, in the real system, the functions are not true sines and cosines, the loop gain in the control systems will vary as the hand rotates, but if the minimum loop gain remains high enough, control will remain accurate over all angles of roll.

A nice demonstration shows an unexpected but realistic effect of these control systems:

- Set the hand up so the elbow is bent a little and the hand is pointing horizontally with the thumb upward (like a karate chop). Now operate the SHOULDER ROLL slider, moving the slider to the left.

The elbow will rotate toward you, but the hand will stay almost stationary. The hand pitch and yaw control systems are compensating for any tendency of the hand to move, and the hand roll control system keeps the hand angle from changing much. You can do this quite easily with your real hand and arm, with about the same accuracy.[34]

Control of a visual relationship.

Let's now add one more layer to this hierarchy.

- Exit Demo 8-2 and choose Demo 8-3, Arm Tracking.

Several of the sliders are gone. They provided reference signals for control systems that are now going to be used by the next level up. When you start the program, a triangle appears, floating in the air. Behind the scenes, three more control systems have been added. Their outputs vary the second-order reference signals for hand elevation, lateral position and reach. Those were varied by the sliders that have been removed.

The three new control systems represent controllers using vision to detect relationships (distances) between the fingertip and the target in the x, y, and z directions.

34 I have since found that some people cannot do this. Try hanging onto a stationary object.

Figure 8-5. The arm is tracking a target with its fingertips. Use the blue-green mouse pad to move the target (the small green triangle) in the grey arm display box.

They act by changing the reference signals for vertical, lateral, and reach positions as you were doing before. Other coordinate systems could be used. As before, we simply posit three perceptual signals corresponding to these three relationship-variables, without trying to guess how the brain might construct such perceptual signals.

Now you can see the blue-green mouse pad (Target X and Y box) with the TARGET Z slider below it. If you click and drag the mouse pointer on the mouse pad, you can move the target laterally and vertically relative to the arm. The TARGET Z slider moves the target toward or away from the shoulder. The arm alters its reach as needed to stay on target. Use the View/Angle controls for different perspectives.

As you move the target, you will see the fingertip track it. Move it any way you wish; the fingertip will follow gracefully (over a reasonable range—this is not a very complex model so it has limitations). The three directions of movement are nearly independent of each other, though the natural directions of movement created at the second level do interact somewhat. You can use the six lateral sliders at the top right of the window to change the hand pitch, yaw, and roll; the finger elevation and lateral position, and adjust the shoulder position with the SHOULDER ROLL slider. Swing the elbow with the SHOULDER ROLL slider. The fingertips will continue to track the target when you move it.

The sliders now remaining relate to the dimensions at the second level that are still free to change, so in using the sliders you are playing the part of more control systems at the third level. along with the three third-level systems that are tracking the target vertically, horizontally, and in the radial direction.

This model is constructed of multiple control systems at each level, and multiple levels. As a model of real human control systems, it's only a conceptual sketch right now. But it shows how the brain's organization could be built up level by level, in a modular way, with lower-level control modules being used for higher-level purposes. Once formed, the modules keep the same organization, and can be used for many different higher control processes, just as human hands can be used for many purposes from grasping a tool to pulling a lever that wins a jackpot or records a citizen's vote. The higher systems do not supersede the lower systems or subsume them: they use the lower systems as their means of action.

Many higher levels of organization can be proposed and tested against real behavior. Because of our ignorance of how the human brain works (most investigations have assumed a very different model and have little to say that would help here), the model we have so far is useful mostly as a jumping-off point for future investigations. I haven't tried to do that kind of work; the point of this book is to explore the principles in this model and try to demonstrate them using interactive programs, so others can carry on from here.

Chapter 9: Demonstrations of Principles

In this last chapter of demonstrations, we will look at some aids to understanding that were developed as various points of difficulty arose.

SQUARING THE CIRCLE

Control of perception means control of the state of an existing perception (as was seen in Chapter 2). It doesn't mean making one kind of perception into another kind. But there is another dimension to understanding control of perception: namely, understanding that we control *perceptions* rather than actions. In Demonstration 9-1 you will be controlling a perceptual variable, tracing a shape by moving a point of light around in two dimensions using the mouse. The mouse movements are your means of controlling the perceptual variable, but the movements themselves are not what is controlled in this task. By the time you have finished with the demonstration, you will know why we speak of controlling perceptions, not actions. Demo 9-1, Square Circle, concerns a perception of a pretty good square.

- Choose Demo 9-1, SquareCircle from the Demonstration Menu.

With the demo open on your screen, you see a very good red square on the screen, and a white dot that is probably moving.

- Use the mouse to make the white dot trace all the way around the square, going very slowly so as to make the tracing as accurate as you can.

You should take about a minute to complete the circuit around the square. The result will be a pretty good white square drawn over the perfect red square (on some computer displays, a perfect square will be a rectangle).

As you will find, this is not an easy task because the connection between the mouse and the white dot is not simple and direct. The dot doesn't simply move the way the mouse moves. One of the points I hope was made in the previous chapter is that reorganizing to achieve good control does not require knowing how the environment works. Therefore, in the interests of pedagogy, I will not explain now why the white dot moves as it does when you move the mouse, and I ask you not to look ahead to the explanation. Just try to accomplish the task, and practice until you can get the dot all the way around with the white trail remaining mostly centered on the boundary line of the square.

- You can restart at any time just by tapping the space bar to clear the trails.
- When you have done as well as you care to do, type v with your free hand, which changes the view immediately.

The square and its white trace vanish, and you see a yellow circle on the screen, with a blue trace wobbling around it in much the same way the white dot moved around the square. This new blue trace shows how you moved the mouse.

Oh, yes it does. You moved the mouse in a pretty good *circle*. Not a square. Here are the results you should expect, when you've practiced as long as I have:

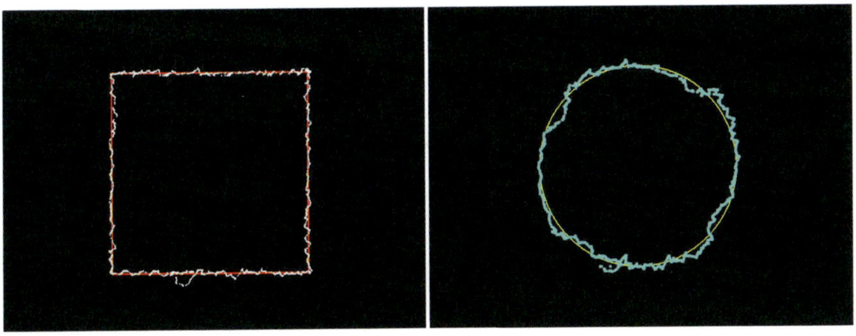

Figure 9-1. Left: White dot controlled by mouse to follow the outline of red square. Right: Blue trace over yellow circle, showing how the mouse actually moved.

I have not yet met anyone doing this task for the first time who was aware that the mouse was moving in a circle. We control the perceived consequences of our actions, not (in most cases) the actions themselves. The normal assumption, of course, is that the mouse must have moved as the white dot it affected was moving. But that is not true.

- With the blue circle still visible, tap the space bar to clear the traces off the screen, leaving only the yellow circle and a blue dot.
- This time, move the blue dot so it traces over the yellow circle as exactly as you can, moving slowly.

This will be much easier since the blue dot moves exactly as the mouse moves. If you make mistakes just tap the space bar to erase the traces and start again. When you have a good blue circle, tap v to change the view and see how well the white dot traced out the red square. Not very well; not well at all. Figure 9-2 shows a typical result.

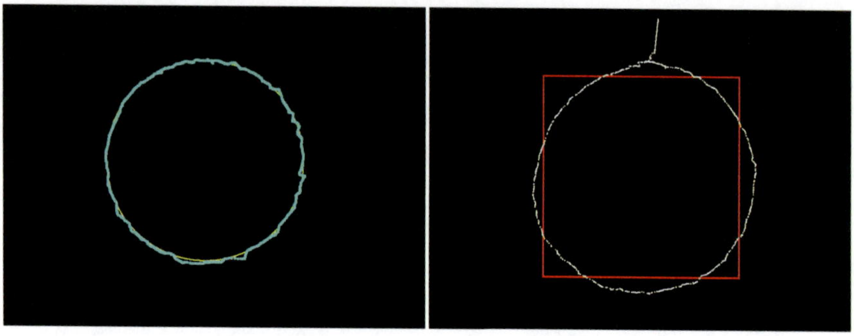

Figure 9-2. Left: Mouse position (and blue dot) controlled to trace over the yellow circle. Right: Resulting movement of white dot over the red square.

Notice that moving the blue dot in a nearly-perfect circle does *not* cause the white dot to move in a square pattern. In fact the white dot moves approximately in a circle, though the ends don't quite meet. You may see even less regular patterns of movement of the white dot, depending on exactly how you move the mouse and where the white dot was when you started. You will never see the white dot tracing out the square, however, although the experimental setup is *exactly* the same whether you are seeing the screen with the square or the screen with the circle in it.

In both of these cases, your movements of the mouse caused the blue dot to move in a pretty good circle (a better one when you could see the circle template). When you were observing the white dot and the red square, the white dot traced the square fairly well. But when you could not see the white dot and red square, the white dot came nowhere close to tracing the square.

The secret lies in the details of the transformation of mouse position into movement of the white dot. In the first figure (9-2) you will notice that the blue dot deviates from the circle in four places that correspond to the corners of the square. These deviations cause the white dot to begin moving radially outward (blue dot outside yellow circle) or inward (blue dot inside yellow circle) at a speed proportional to the amount of deviation. To make the white dot deviate from a circular motion enough to stay on the square as the dot moves around it, *it is necessary for the blue dot to deviate slightly to the outside or the inside of the yellow circle.* And you will see that it does, even though you had no way of seeing that happening and did not know you were doing it. Your control system responded to tracing errors by making the correct movements, but you did not perceive that. And of course, even if you try to notice what your hand is doing, you will need a lot of imagination to believe you can perceive it moving in a circle while you're trying to control the white dot.

The point of this demonstration is simply to show that you control perceptions by varying, not controlling, the actions needed to carry out the controlling. When you hear the term "control of perception," think of squaring the circle.

That experiment can be repeated using a triangle instead of a circle as the template for the movement. Just type 'T' or 't' to use the triangle (and 'C' or 'c' to go back to the circle). This is considerably harder and you have to move more slowly to come close to succeeding. When you have made the white dot stay reasonably close to the square and look at the alternate view, you will see that the blue dot traces out a triangle, with a yellow triangle behind it. All the preceding variations, and those that follow, can be done using the triangle template exactly as you did with the circle. The following discussion assumes that the circle is being used.

For those who want to go into this more deeply there are several additional commands you can use to change the experiment. The first thing you can do is to show both views at the same time.

- Type 'B' or 'b', and you will see them Both, superimposed.
- Type space to clear the traces, and watch either the white dot or the blue dot.

If you make the blue dot follow the yellow circle, the white dot may show some other behavior but it will not trace the red square. But by watching the white dot you can make it trace the red square, and then the blue dot will still stay fairly close to the yellow circle. You can see how the transformation works by experimenting.

If you type 'D' (or 'd'), the transformation will change to a Direct effect rather than one with radial speed of movement as a factor (type D again to return to the more difficult transformation). The word direct will appear in the center of the display. Now the radial error between the blue dot and the circle is simply added to the radial distance from the center to the square, to position the white dot. When you reduce the radial error between the white dot and the square to zero, the white dot is on the square and, necessarily, the blue dot is also on the circle. Of course this works equally well the other way: if you control the blue dot to be on the circle, the white dot will be on the square.

- Type B again to return to the individual views.

From the individual views, you can repeat the first two experiments. This time you will find that making the blue dot stay on the circle results in the white dot tracing out the square, and vice versa. Now it works both ways. Perhaps this will show that it is easy to be fooled: just because A corresponds to B, you may be tempted to think that B, therefore, corresponds to A—which is certainly not the case in the first experiment, though it is in the second one.

THE CROWD (OR GATHERING)

The next demonstration is one of the oldest; I wrote the first version in the 1980s for Professor Clark McPhail of the University of Illinois and Professor Charles Tucker of the University of South Carolina, both sociologists. McPhail and Tucker were interested in the idea of social laws being emergent from interactions of individuals. McPhail in particular had studied the behavior of crowds all over the world for many years, and wanted to know how crowds or gatherings of control systems would behave. The purpose of this demonstration was to show how social-looking patterns can emerge from interactions in a group of individual control systems having fairly simple properties.

The demonstration involves a white field in which small circles are displayed. The black circles that stay in one place are obstacles.

There is usually one larger stationary destination circle, green. I will call the other circles, the ones that can move, people. A person who seeks a destination position is colored red; one who seeks the position of another person is colored green (a smaller circle than the destination).

A person in this demonstration has three possible kinds of two-dimensional control systems. One kind avoids collisions with obstacles or other people. Another kind seeks a destination position at some other place on the screen. The third kind tries to establish a relationship to another person, following that person so as to maintain a specific distance from that person, and a specific angle of sight to that person relative to the direction of travel ("following" means keeping the followed person at a bearing angle of zero—straight ahead). All these control systems operate at the same time, though usually only two of them are used at once in one individual, the collision-avoidance system and one of the others.

The collision-avoidance system uses signals from two simulated *proximity detectors*, one detecting the sum of proximities of all objects on the left of the direction of travel, and the other detecting the sum on the right. A *cardioid sensitivity pattern* provides the highest sensitivity to objects straight ahead, and zero sensitivity to objects directly behind. Proximity is defined by an inverse-square law, decreasing as the square of the distance to an object increases, just as the apparent brightness of a light decreases, and having a maximum value of 255 when the distance is zero. All objects contribute to total proximity, but the nearest ones contribute by far the most to the proximity signals on the left or the right. The other two kinds of control systems also use proximity detection, but in different ways.

The user of this program can design many different setups, several of which are included as demonstration files which this program can use. Instructions for creating new setups are given later.

Operating the program.

> *Tip: Before you start, it's important to know that in some of the demonstrations that follow you'll have a chance to save your working file. If you want to keep the original stored files so you can reuse them, be sure to use Save As and rename your working file.*

- Start Demo 9-2, Crowd.

The Crowd demonstration opens with a window showing some preprogrammed setups for this demonstration. Select the demonstration file called single.per from the list that appears in a dialog window. Either double-left-click the name, or left-click once and then left-click open.

- Click RUN.

A single person appears on the screen as a red circle at the upper left, and a destination appears as a (larger) green circle at the lower right. The person moves among stationary obstacles and eventually reaches the green destination circle.

- Click the TURN TRAILS OFF toggle button to change it to turn trails on. (A red trail is left behind the moving person.)

This demonstration is most interesting with the trails turned on. A run can be restarted at any time with the RUN button. A new set of randomly-placed obstacles is generated each time. After you have clicked on a button, tapping the space bar has the same effect as the click. This program has proven to be a bit addictive.

Figure 9-3 shows the screen after the moving person has reached the destination, with trails turned on.

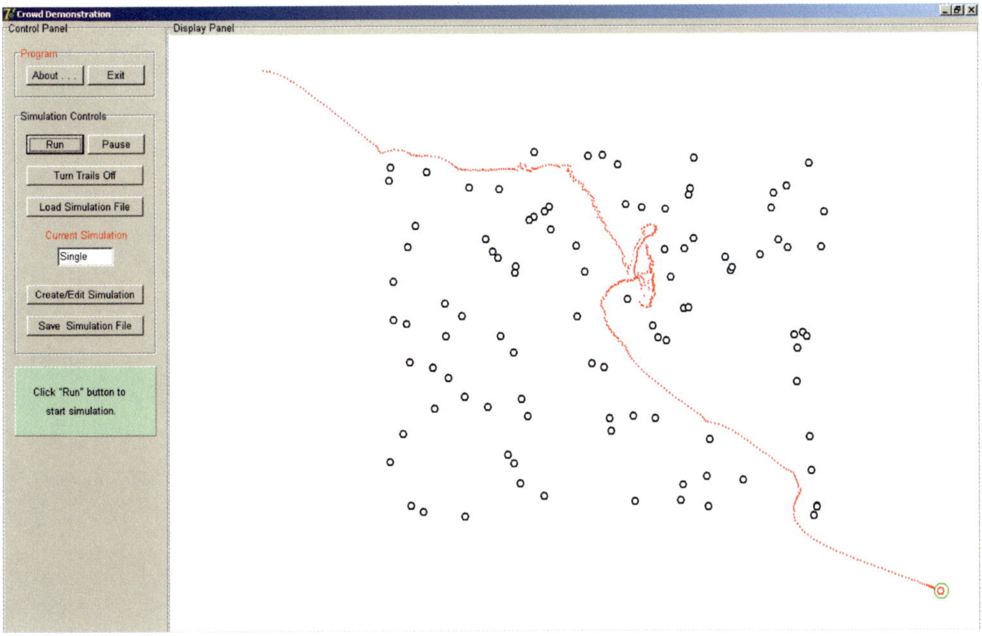

Figure 9-3. Path of single person through field of obstacles, with Trails turned on.

Remember, once you have clicked the RUN button, you can generate new runs by tapping the space bar. With each new run, a new set of obstacle positions is generated at random.

Note that when the person approaches a dense array of obstacles, the path actually turns away from the destination circle, and even reverses. A psychologist seeing this demonstration for the first time exclaimed "It's intelligent!" He explained that in psychology, one sign of intelligence has been that a goal-seeking organism can turn away from the goal in order to find a different path that gets it out of a trap. It was assumed that this kind of behavior requires abstract reasoning, the ability to survey and analyze spatial relationships, and other such admirable capacities. The trail left by this person moving among the obstacles will frequently show this phenomenon—in fact once in a while the person will seem to be hopelessly trapped, moving again and again through the same pattern, but in most cases will eventually break free and find a successful path. It seems to act much like “The Little Engine That Could,” and is sometimes cheered by onlookers when its persistence is rewarded.

The Crowd demonstration shows how we can mistake complexity of behavior for intelligence. I’m not disputing the existence of intelligence (though I think that’s a loose and uninformative term). I mean only that one can make overly complex hypotheses about how an organism accomplishes what it is doing, when the actual mechanism is much simpler. In the case of the Crowd demo, the impression of intelligence is quite misleading, for the person is doing none of the things one might read into the behavior. I know that is true because I wrote the program. The person is seeking a destination position and avoiding collisions, and that is all. No decisions are being made, there is no analysis of possible paths through the field of obstacles, no planning of trajectories, no remembering of past experiences, no searching for a way out of a trap. The behavior we see is determined completely by the actions and interactions of two simple control processes in an environment containing randomly placed obstacles.

Mars Spirit and Opportunity Rover programmers take note.

The Guru setup.

This one is best viewed with the trails off. Click LOAD SIMULATION file and select guru.per (the current run will end when you do that). When you click on RUN this time, you see a red circle on the left side moving among obstacles as in the previous demonstration, seeking a goal position on the right side. A group of green circles appears at the lower left. Each of these green circles contains two control processes: one of them avoids collisions with other objects, moving or stationary, and the other tries to maintain a constant distance from the red object while aiming directly toward that object, the “guru.” The destination-seeking control systems are turned off in these objects, though you can try turning them on to see what happens.

When McPhail, the crowd expert, first saw this demonstration, he said "Arcs and rings! Arcs and rings!" Apparently this is a frequently-observed phenomenon in gatherings where there is some common focus of interest. People gather around the focus, "jockeying for position" as the saying goes, ending up in an arc or, with more people, a ring around the object of interest. As you see, the green followers of the guru do exactly that. But they do it only because each one separately seeks a particular relationship to the guru, while avoiding too much proximity to obstacles or other people. There is no explicit social law in this program that says arcs and rings must form. The interactions among the control systems just generate that result.

The Konrad setup.

Click LOAD SIMULATION FILE and select Konrad.per. This demo starts out much like guru, but quickly shows a different result. Here, one of the green followers attempts to maintain a particular spatial relationship to the red circle representing Konrad Lorenz. The second green follower attempts to maintain the same spatial relationship, but not to Konrad; it follows the first green circle. The third follows the second, the fourth follows the third, and the fifth follows the fourth.

The result looks just like the goslings who were said to have imprinted on Konrad Lorenz, following him all around the barnyard just as they follow a mother goose. But you will notice that this is not the correct description. If each gosling had learned to follow Konrad, we would see the result in the guru setup. To get a string of goslings all moving in one line, we have to make all the goslings but one imprint on another gosling instead of Konrad. It would be interesting to see a replication of Lorenz's demonstration of imprinting in which the barnyard contains a scattering of obstacles and a video is made of how the goslings move. One wonders if it would be possible to get the first gosling to imprint on the last one. Also, one wonders if the goslings always end up in the same order. Might the simple avoidance of too much proximity make goslings seeking the position of other goslings move to the end of the line? Something similar to that happened in the guru setup as the people jockeyed into their final positions.

The Rush Hour setup.

In this last setup, we have two mobs of 20 people each seeking different destinations. Click LOAD SIMULATION file and select RushHour.per. One mob starts at the upper left and seeks a destination at the lower right. The other mob starts lower right and seeks the upper left destination. Each individual seeks a destination, while avoiding collisions with other people. No obstacles are used, though you can add some if you wish. The initial positions of the people are randomized around the initial starting points.

The result is a little like what happens on a city's sidewalks at lunch hour, except that the people here are not as smart as the real ones, and can't reorganize. The two mobs start toward their respective destinations until they meet. Then we see a melee as each person-object tries to make its way forward while avoiding collisions. You will often see two people locked in a conflict when they meet almost head-on, moving off in some irrelevant direction until they finally work their way free and make their belated ways behind their brethren. The Walt Disney cliché of the last one in the group scurrying to catch up happens almost every time because of course there is always a last one.

This demo demonstrates the title of one of McPhail's books: *The Myth of the Madding Crowd* (McPhail 1990). What we are seeing is the behavior of two mobs, two crowds. But mobs and crowds do not behave; only individuals do. What we are seeing is strictly the result of interactions of individuals with everyone and everything else around them, in relation to their individual goals. There are no laws operating at the crowd level, or if there are, they are completely a result of the laws operating in each individual (plus, of course, the laws of geometry and physics).

Editing parameters.

You can customize your own versions of the Crowd demo by editing the program parameters.

- Click CREATE/EDIT SIMULATION to see the screen in Figure 9-4.

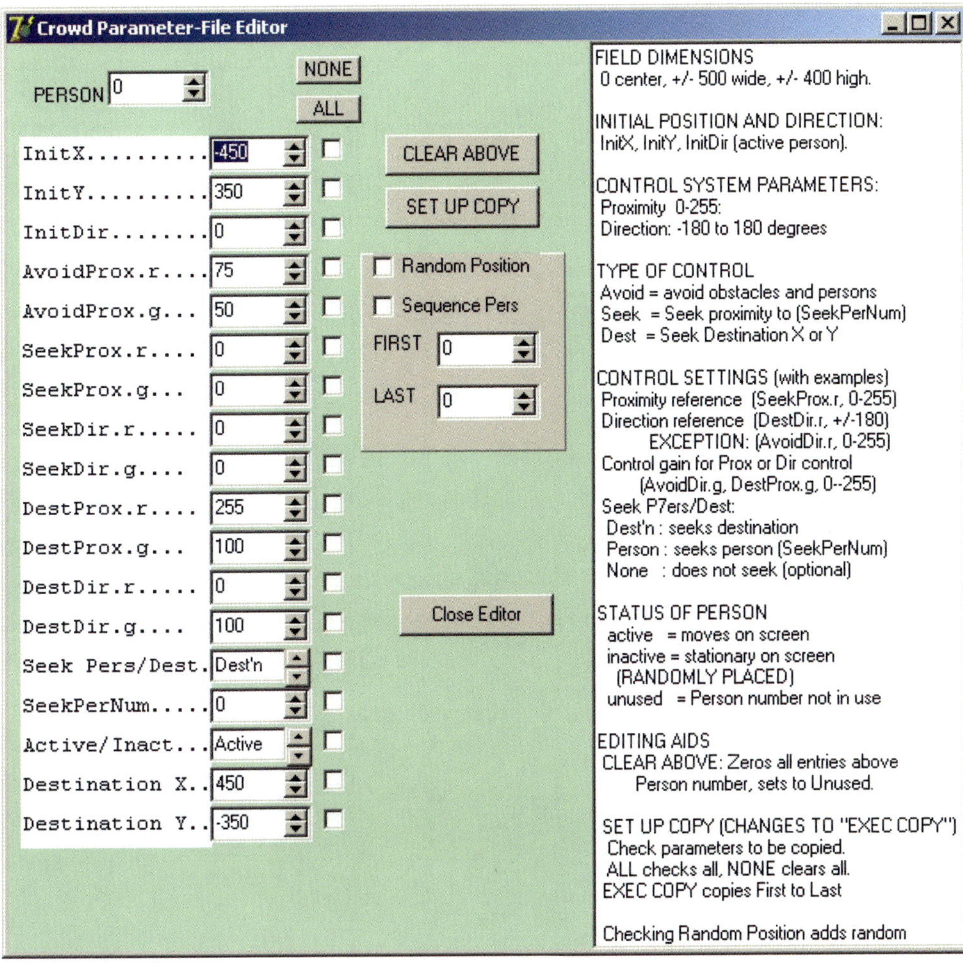

Figure 9-4. Editing parameters screen.

In the left column all the parameters that can be adjusted for each person are listed. The initial X and Y positions and the initial direction of travel (in degrees) are shown, followed by ten parameters for controlling proximity and direction relative to other objects or the destination (more discussion farther below). The next three entries allow selecting whether the "seek" control systems seek a person's position or the destination and if seeking a person, what person number to seek. Then comes a four-way selection making the object **unused** (not present or shown at all), **inactive** (present but stationary on the screen in some randomly selected position that changes with every run), **fixed** (randomly selected the first time but not changed from run to run), or **active** (control systems working, object moves on screen). Finally, if the object seeks a destination, the destination's X and Y coordinates can be selected. Notice that at the top, the person number is shown as zero—this is the first of 256 possible objects numbered 0 to 255. The parameters shown are for the one active object in the "single" demonstration.

Each entry can be changed by using the up-down triangles at the right of each spinner box. Numerical entries can also be typed in directly.

For complicated setups is it useful to be able to create copies of selected entries, or all of them. The check boxes mark which entries are to be copied, with ALL and NONE buttons at the top to save time. The CLEAR ABOVE button simply sets all parameters to zero for person numbers *greater than* the number shown after PERSON. All those persons are made unused.

The last button is SET UP COPY. Click to switch it to EXEC COPY. You can specify several more directives in the area just below, as shown in Figure 9-5:

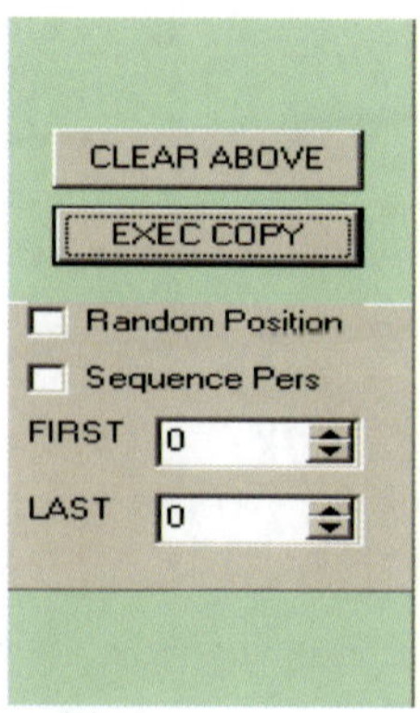

Figure 9-5. SET UP COPY toggles to EXEC COPY.

After the parameter check boxes have been set up (either before or after clicking on SET UP COPY), you can do the copying with several modifications that save time (individual entries can always be manually edited). Fill in the range of person numbers, FIRST TO LAST, that are to receive the copies.

When copying to a group of consecutive person numbers, it is usually desirable not to have all the initial positions in the same place. To save manually changing each set of X and Y coordinates, you can check the Random Position box, and the initial position for both InitX and InitY will be the position shown in the displayed values plus random numbers added to both X and Y.

For cases in which a series of people are to seek the position of a person with an earlier number (as in the Konrad setup) checking the Sequence Pers box will, for each person number in the range, fill in the seek-person number entry to a value one lower. Person N will seek the position of person N-1. The first person in the range seeks a person number one *below* the range.

- Finally, click EXEC COPY to execute copying.

This copying goes into the set of parameters in working memory that are used by the program for the setup currently in effect. *The parameters in the setup stored on disk are not changed. The only way to change them is to click* SAVE SIMULATION FILE *before exiting the program and before changing to* CREATE/EDIT SIMULATION. *Saving can be done at any time. If you don't want to keep an unsuccessful setup, just don't save the parameters.* When you click SAVE SIMULATION FILE be sure to give it a new name for this new setup, or just save it in place of an existing one if you don't want to reuse it.

While there is no required order for entering data for different kinds of objects, it is generally most convenient to enter data for active destination-seeking objects first, then for active person-seeking objects, and finally for inactive or fixed objects that serve only as obstacles. This makes it easy to change the number of obstacles. The control parameters will most likely be the same or similar for active groups. To enter a group of objects with similar control parameters, set PERSON to the first person number in the group, enter data for the that one manually, then copy it into a range of person numbers starting one higher than the person number. Be sure to put checks in the boxes indicating which parameters are to be copied—if you leave all the boxes unchecked, nothing will be copied.

When you finish editing, click CLOSE EDITOR. You can immediately RUN the current setup, then go back to editing, and so on until you're satisfied with the result. When you're satisfied, DON'T FORGET TO CLICK SAVE SIMULATION FILE OR ALL YOUR WORK WILL BE WASTED—IT WON'T BE STORED ON THE DISK. Heed this caution before exiting the program or opening a new setup file.

Editing control parameters.

The basic characteristics of the active objects are determined by setting the parameters of the control systems that exist in each one. In the following pages, you can skip the paragraphs showing program source code if that is of no interest—but do notice how little of it there is. Feedback control is conceptually simple.

This is not a general-purpose programming language; while you can do some things, not everything you can think of is possible. You can blame the author if something you try doesn't work as you wanted it to. However, I hope that you or someone else will then get to work and correct the deficiencies in the program.

Collision-avoidance (avoidprox.r, avoidprox.g).

Avoidprox.r determines the threshold amount of total proximity, left plus right, at which the person will begin to slow down and turn either left or right. Avoidprox.g is the «gain» control which determines how rapidly the turning will take place after the threshold proximity is exceeded.

If avoidprox.r is set to 100, the turn to the left or right will begin at distance somewhere between 0.01 and 0.02 of the screen width, or roughly 10 to 20 pixels. If there are multiple nearby obstacles dead ahead, the turn will begin sooner. Avoidprox.g determines how rapidly the turn will take place.

Collision-avoidance actually combines two control systems, one acting when the direction of most proximity is to the left, the other when it is to the right. Some random dither is added to the output of this control system to prevent its getting hung up in repeated patterns. After the left and right proximities are computed, this is the entire code for avoiding collisions:

```
with AvoidProx do
begin
    p := RightP + LeftP; // Left and Right proximity
    e := r - p;
    if e > 0 then e := 0;
    if LeftP < RightP then e := -e;
    o := 0.02*g*e*(1 + (random - 0.5)); // 50% dither
end;
```

The absolute value of the error signal e is also used to slow the speed of the object when the proximity is larger than the reference amount. The output o contributes to the speed of left or right turning.

Controlling proximity to another person: seekprox.r; seekprox.g.

This control system works by adjusting the speed of travel. It reduces the speed when the object is too close, increases it when the object is too far (relative to the reference proximity). The perceptual signal in this system is the line-of-sight proximity (RelProx) to the position of the person selected to follow. The output signal o contributes to the speed of travel (speed of travel is also affected by collision avoidance and distance to the goal when those control systems are activated).

Here is the code:

```
with SeekProx do
begin
    p := RelProx;
    e := r - p;
    o := 0.01*g*e;
end;
```

Controlling direction to another person relative to direction of travel: seekdir.r, seekdir.g.

This system detects RelDir, the angular direction of the line of sight to the selected person, relative to the direction of travel. It compares that perceived direction angle, p, to the reference direction angle, r. The output signal contributes to the rate of turn to left or right. Setting r to 0 means to travel so the sought object remains straight ahead. Here is the code:

```
with SeekDir do
begin
    p := RelDir;
    e := r - p;
    while e > pi do e := e - 2*pi; // bring angle in range
    while e < -pi do e := e + 2*pi;
    o := -0.2*g*e;
end;
```

Note that angles entered in degrees are converted to radians for use here. The error in angle, if outside the range of -180 to 180 degrees because of multiple turns, is brought within the proper range.

Seeking the destination: destprox.r, destprox.g, destdir.r, destdir.g

The third set of control processes detects the direction to the destination relative to the line of sight as well as the distance to the destination. The proximity reference level destprox.r is usually set to the maximum possible, 255, so the person will stop in the center of the goal circle. The direction reference signal, destdir.r, is normally set to 0 so the person will try to head directly toward the goal. The two gain signals are adjusted for the best approach to the goal. If set too low the person can end up circling around the goal. If set too high, the urge to get to the destination can overcome the collision avoidance systems. The perceptual signals are computed earlier in this procedure rather than at the time the control behavior is computed, so the code for these two systems looks like this:

```
with DestDir do
begin
e := r - p;
while e > pi do e := e - 2*pi; // bring angle in range
while e < -pi do e := e + 2*pi;
o :=  -0.2*g*e;
end;

with DestProx do
begin
e := r - p;
o := 0.002*g*e;
end;
```

The velocity is limited to prevent ultra-fast movement when the person is far from the destination.

Adjusting the parameters and seeing their effects is easy. On the Crowd Demo screen, click EDIT PAR to open the editing window, change one or more parameters, click END EDIT, and then on run. No changes are saved permanently until you click on SAVE PAR, so you can make as many adjustments as you like until you have an acceptable setup. You have a chance to give a name to the setup after you click SAVE PAR. (I trust that these repetitions are more useful than annoying.)

THE INVERTED PENDULUM

The last demonstration, first shown at an international CSG meeting in Germany in 1994, illustrates an interesting principle of hierarchical control. It addresses a control problem that is often given to control engineering students. Imagine a movable cart from which a pendulum is suspended—perhaps the tracks on which the cart runs are elevated, so the pendulum, hanging off one side of the cart, can swing freely below as well as above it. A motor is used to apply a force to the cart, and in this version we assume that the cart is equipped with sensors that report its acceleration and velocity. In addition, there are sensors that report the angle of the pendulum where it is pivoted, the angular velocity, and the position of the bob in space (in the X direction).

The object of this exercise is to design a control system that acts by applying a variable force to the cart, to control the position of the bob so it is balanced on top of the arm of the pendulum. Richard Kennaway has done an extensive analysis of this system and reports it in the Appendix at the end of this book, so I will skip the details.

- Before going further, start Demo 9-3, The Inverted Pendulum.

You will see this screen. The labeled buttons give you a pretty good idea of what you can do with this program.

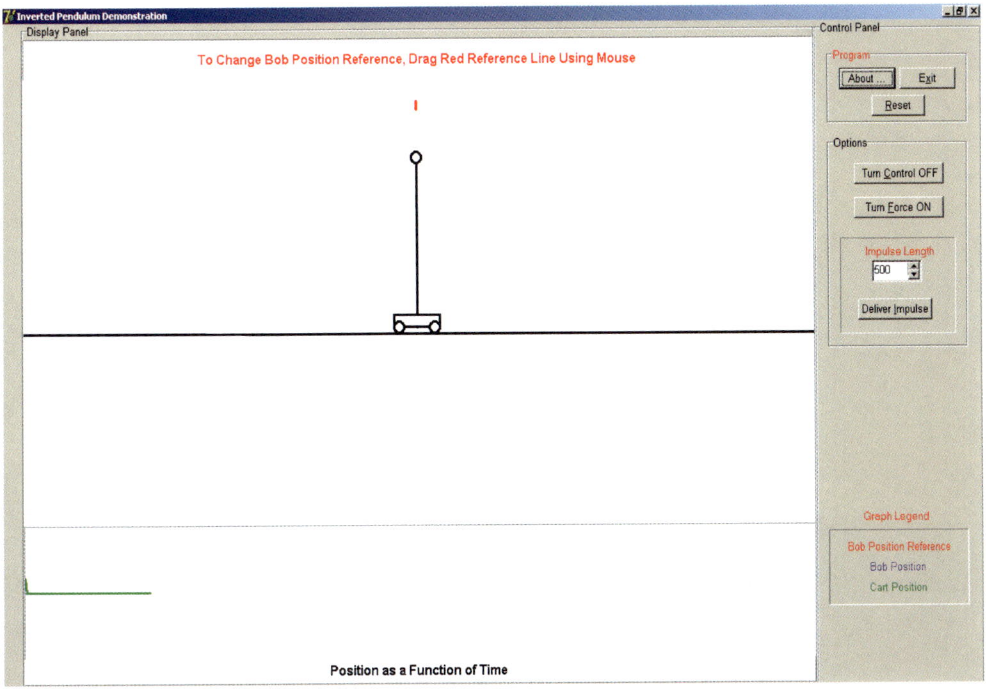

Figure 9-6. The Inverted Pendulum at startup.

To make something happen, click and drag sideways on the red mark just above the pendulum bob. You will see the cart move in such a way as to start the pendulum bob falling toward the new position of the red target mark. Then the cart moves ahead to decelerate the bob, finally letting the bob catch up just in time to bring it to a stop exactly beneath the mark.

If you click TURN CONTROL OFF, eventually the bob will start swinging freely, its forces moving the cart, too. Turn control back on when the bob gets to the top of its swing, and control will be restored. This system isn't quite smart enough to achieve balance when the bob is hanging downward or very far from vertical, so you have to restart the program if that happens.

You can apply a steady force or an impulse force to the bob, and see how the cart moves to counteract the effects. For the steady force, the bob will come to a position where the pendulum is leaning into the applied force (as into a wind) just far enough to maintain equilibrium—even when the reference position is changed.

This result is achieved using a total of four hierarchically-related control systems of very simple design. The lowest level system controls the velocity of the cart by varying a force applied to it. The second level controls the position of the cart by varying the reference signal sent to the velocity control system. The position of the cart and the current position of the bob, together with gravity, determine the forces applied to the bob to accelerate it. The velocity of the bob is controlled by varying the reference level for cart position, and the position of the bob is controlled by varying the reference level for its velocity. The reference position of the bob is set by moving the red mark.

Each control system is a simple proportional controller, stabilized by the fact that its feedback connection contains a single integration. Richard Kennaway's analysis gives more details. The appropriate movements of the cart follow from this design; they are not explicitly programmed. As in the case of the crowd program, we see what appear to be very complex cognitive decisions being made when the truth is that only a few very simple control systems are involved.

A system organized this way *must* behave as we see it doing.

This controller of an inverted pendulum obviously must have some application to the problem of modeling bipedal walking. At each step, the body is moved forward to match a moving reference position, and only a few hierarchically-related control systems are needed for each leg to maintain it in balance—in addition to the systems that sequence the control back and forth from leg to leg and keep the body upright. This is another of many future projects which reside tantalizingly on the dusty shelves waiting for time and resources, as well as energy and youth which are in short supply just now.

With that we have reached the end of this book, but for a few closing remarks. I hope you have had an adventure worth the time you have spent on it.

Chapter 10: The Purposive Self-Motivated System

I learned to write digital computer programs in the mid 1950s, on an IBM 650 in a back room of the Dearborn Observatory where I had, a few years previously, worked and lived as an undergraduate at Northwestern University. This kind of programming came into being after I had already learned to program an analog computer, the sort of computer (begat by the slide rule out of Napier's Bones) that flourished briefly before the digital revolution. Ever since then, whenever I had the opportunity, I have been tricking digital computers into behaving like analog computers, with the results you have seen in this book.

Why "tricking?" Because it is the analog computer, not the digital computer, that is organized the way living systems are organized. Digital computers with operating systems such as Windows or Unix are viewed and used as "event-driven" systems. They do nothing until some external stimulus, such as a user clicking on a button on the screen, stirs them into action. Then one event leads to another, just as the behaviorists at the turn of the last century and the biologists before them—and many, many scientists after them to this very day—believed organisms to work. When all the events have played themselves out, some data are stored or printed out or sent outward on the internet, and the whole system goes inert, waiting for the next event to bring it to life again.

Digital computers are programmed by writing instructions in excruciating detail telling them what to do. If this, do that, otherwise, do the other thing. But analog computers are programmed simply by assembling and plugging in their parts, connecting the parts together with patch cords, and turning them on. The analog computer behaves as it does because of the physical properties of its components and the relationships among them, all the parts operating continuously and simultaneously. What the analog computer does arises spontaneously out of its physical and organizational nature, not out of prior events. Simply supplying the analog computer with a steady source of electrical power is enough for it to begin producing highly-organized patterned behavior.

In all the demonstrations you have seen in this book there is a "heartbeat," a simple procedure that comes to life (usually 60 times per second), triggered by a clocking circuit built into the computer chip. That determines the basic "dt," or interval of time, that allows the programs to initiate their own behavior, seek their own goals, control their own perceptions. The digital nature of the computer is smoothed out by computations, so that as far as the models are concerned, time is a continuous flow and many different processes are occurring at the same time—just as in analog computers where all computing components are active literally at the same time, and just as in organisms where all cells are working continuously, not waiting to start functioning until some other cell is finished with its business. Within a single dt, all the perceptual signals change slightly, and all the reference signals and error signals and output quantities and input quantities and disturbances, just as if there were a lot of simple independent parts operating in parallel, and the computer were sampling all the continuous processes 60 times per second for us to look at. The digital computer, poor thing, can do only one operation at a time, but we trick it into acting as if it can do many operations at once. As if it were a living system.

Herein lies the Copernican revolution that is Perceptual Control Theory. No longer can we see organisms as sensitive responders to discrete stimuli, behaving as they are shaped to behave by the environment. Instead we must see them as active agents born with their own deep inner purposes and capable of altering their own organizations to control the world around them in service of those purposes. We must see species not as the products of selection pressures that randomly change and capriciously winnow out organisms, but as the result of inner reorganizations that change what matters the least to the organisms in order to control what matters the most to them. Living systems, while obeying all the laws of physics and chemistry as must everything in this finite universe, create and then obey organizational laws that end up dominating the physical world with which organisms interact, imposing order where it would not have existed but for life. The world with living systems in it is very different from the world that would exist without them.

In the end, we will give back some of the digital-style processes to our models. Organisms like us do in fact compute with symbols and rules, using discrete either-or logic as well as continuous analog computations. The PCT model, in its hierarchical form, has placeholders for such digital processes, though they will probably turn out to be a hybrid with both digital and analog properties. But never again will we be able to support the notion that organisms behave simply as a consequence of what the environment does to them. The behavior of organisms is not event-driven. It is the ultimate driver of events, as well as all the ongoing continuous processes that together amount to Life Itself.

Epilogue: What Good Is It?

If behavior is governed by control systems at every level from top to bottom, if the fundamental process of life is negative feedback control, it quite possibly follows that all other theories about how people and other organisms work are wrong. I think that is basically why it is taking so long for the life sciences to get around to adopting control theory. Not only must all the wrong ideas be cleared away and all the conclusions drawn from them be canceled, but the people best equipped to accomplish the housecleaning are the very ones who have the most to lose from it. For every wrong idea and every false conclusion, a hero is required who was at one time firmly convinced that the idea was right, whose reputation was built on its rightness, who has earned a living from its rightness, who has taught innocent youth about its rightness. One can understand why it's hard to find such heroes, but I have met a few. The Control Systems Group contains a number of them.

When Queen Victoria asked Faraday "What good is electricity?" he allegedly replied "What good is a baby?" What good is perceptual control theory? I would like to find out. I hope that more heroes volunteer pretty soon.

Again and again I have been backed into a conclusion, and I hope many readers of this book find themselves in the same position. My late wife Mary said it best (as she often did): she said that people are not like control systems, for the same reason that the heart is not like a pump. The heart is a *pump*, she said, because it pumps. She stopped there and waited for me to get the point and say: People *are control systems*, because they control.

Mary also said, referring to these computer models in her typical indirect way, "At the level of blah, blah, blah, perceptual control theory is no better than any other theory of behavior. The models make the difference." She meant, I think, that purely verbal descriptions lack rigor, though she would never have uttered such a stilted sentence. I hope this book has taken us a notch above the level of blah, blah, blah.

Lafayette, Colorado

November, 2008

AFTERWORD

A Possible Confusion

As Richard Kennaway points out in his Appendix, Modern Control Theory (used as a model of living control systems) assumes that there is a model of the environment inside the control system—inside the brain—while PCT does not. We use models of the brain and its environment in this book. However, the part of the model that represents the brain does not contain any model of the environment. If it were an MCT model, there would be two models of the environment in the diagram, one shown as existing inside the brain, and the other representing the actual environment. If you look up model-based control on the web, you will see diagrams like this, with one model of the environment (often called "the plant") outside the controller, and another model of the same environment, the same plant, inside the controller. It is assumed that the model of the environment shown outside the controller is absolutely accurate, while the model inside it is, for some reason, less accurate—even though the model shown as outside the brain is the product of somebody's brain (though it's sometimes hard to see what part of the diagram refers to what is outside the controller).

If you don't believe this, look up http://lorien.ncl.ac.uk/ming/advcontrl/sect3.htm.

You and I use the model-perceiving levels of our brains (the "program" level, as defined in Powers, 1989) as we explore what is in this book. It has taken me a whole book to communicate a partial understanding of negative feedback control. I would not dare, as we labor to construct this model of a human being, to venture to construct a working model of how my proposed ninth level of organization of the brain does its thing. It doesn't seem that the brain's model-building capacities work fast enough to accomplish very much in real time. At least mine don't work very fast—I've been working up to writing this book since the 1970s.

I don't dispute the claim that brains, at some level of organization, contain models of the environment. I do dispute the claim that this level of organization can control anything very fast or very accurately, in comparison to the kinds of negative feedback control that I propose to exist at the lower levels of organization, the kinds explored in this book. Perhaps the two points of view will one day be reconciled, as belonging to different levels of discourse about the same thing. But that day is not today.

References

Powers, William T. (1989); Making Sense of Behavior. p. 135 ff. (Bloomfield, New Jersey, Benchmark Publications)

I have not tried to make this a scholarly book. I haven't the resources. But a lead into the literature of perceptual control theory can be obtained from two sources:

First, Dag Forssell's ongoing project to provide PCT resources online:

http://www.livingcontrolsystems.com

Second, the Control Systems Group's web page:

http://www.perceptualcontroltheory.org

And then there is always Google.

Appendix

In this appendix we present some mathematical analysis of the control systems described in this book. We assume familiarity only with basic algebra and linear differential equations. In places we will use Laplace transforms to simplify the solution of linear differential equations in multiple variables. A good reference for the control theory used in this book is [Lei04].

1 THE PROPORTIONAL CONTROLLER

We shall begin with the control system demonstrated in Chapter 3, pictured in Figure 1.

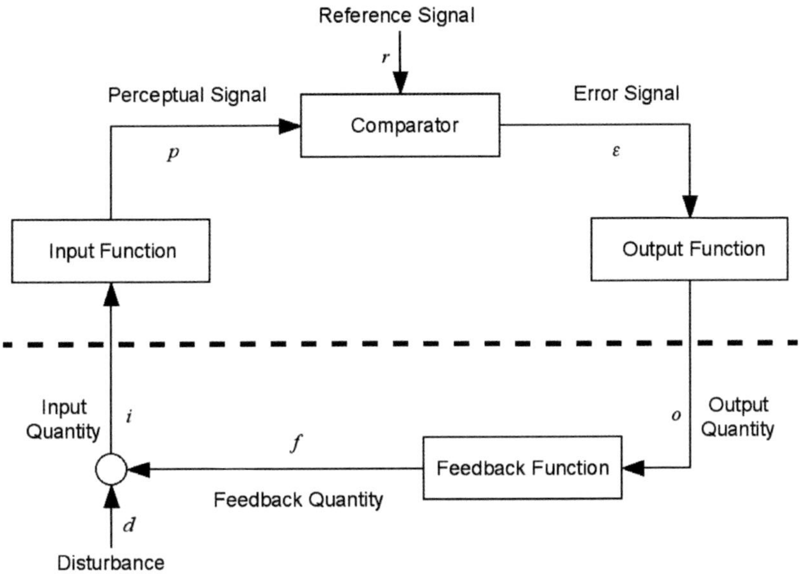

Figure 1: Control system

Here are the differential equations which describe the system. We use ϵ rather than e for the error to avoid confusion with e, the base of natural logarithms.

$$\begin{aligned} \epsilon &= r - p \\ \dot{o} &= (g_o\epsilon - o)/t_o \\ f &= g_f o \\ i &= d + f \\ p &= g_i i_{t-t_i} \end{aligned} \tag{1}$$

r and d are external signals, and hence arbitrary functions of time. The time history of all the other signals is determined by the above equations from r and d. We therefore wish to solve the equations and express each of the quantities ϵ, o, f, i, and p in terms of r, d, and the system parameters g_o, g_f, g_i, t_o, and t_i. We can then directly observe how the quantities vary when the external signals and system parameters change.

1.1 The steady state

First, we shall consider the steady state behaviour of the system, assuming for the moment that it is stable. (For some values of the parameters, we shall see later that it is not stable.)

In the steady state, with all signals constant for all time, $\dot{o}$ is zero, and $i_{t-t_i} = i_t$. The equations thus reduce to:

$$\begin{aligned} \epsilon &= r - p \\ o &= g_o\epsilon \\ f &= g_f o \\ i &= d + f \\ p &= g_i i \end{aligned}$$

These are easily solved to express each quantity as a linear combination of r and d, most clearly presented as the following table. For example, the first line means $\epsilon = r/(1 + g_{oif}) - dg_i/(1 + g_{oif})$. We abbreviate products of gains such as $g_o g_i g_f$ to g_{oif}.

	r	d
ϵ	$1/(1 + g_{oif})$	$-g_i/(1 + g_{oif})$
o	$g_o/(1 + g_{oif})$	$-g_{oi}/(1 + g_{oif})$
f	$g_{of}/(1 + g_{oif})$	$-g_{oif}/(1 + g_{oif})$
i	$g_{of}/(1 + g_{oif})$	$1/(1 + g_{oif})$
p	$g_{oif}/(1 + g_{oif})$	$g_i/(1 + g_{oif})$

The control system is set up in Chapter 3 with g_o very much larger than g_i and

g_f. When this is the case, the table can be approximated as:

	r	d
ϵ	small	small
o	close to $1/g_{if}$	close to $-1/g_f$
f	close to $1/g_i$	close to -1
i	close to $1/g_i$	small
p	close to 1	small

Thus the perception tracks the reference almost exactly, and is almost unaffected by the disturbance. The input, by definition, does the same as the perception, scaled by $1/g_i$. The feedback is almost exactly equal and opposite to the disturbance, and is hardly affected by the reference. The output varies significantly with both the reference and the disturbance, and the error varies hardly at all. This is how we expect a well-functioning control system to behave.

1.2 Transient behaviour, with no input delay

To determine when the system tends to the steady state, and how fast, we shall first consider a simpler version, in which there is no input delay ($t_i = 0$). We shall later consider the effect of a positive input delay.

It is most straightforward to eliminate all variables but o from the equations (1), to obtain the following differential equation. We write k_o for $1/t_o$.

$$\dot{o} \quad = \quad k_o g_o r - k_o g_{if} d - o k_o (1 + g_{oif})$$

If r and d are constant, the solutions to this exponentially approach the steady-state value previously computed, with a time constant of $t_o/(1 + g_{oif})$. The other quantities in the equations all approach their steady states with the same time constant. When the loop gain g_{oif} is large, it is the amount by which the open-loop time constant t_o of the output block has been speeded up.

Thus the system is stable for all positive values of the gains and output damping. By making the output gain large enough, we can make the controller reach its steady state as fast as we like, irrespective of the open-loop time constant of the output block.

1.3 Input delay

The story changes when we reintroduce the input delay t_i. We find that an important parameter of the system is the ratio $\gamma = g_{oif} t_i / t_o$. The system turns out to be unstable, exhibiting exponentially growing oscillations, if and only if γ is greater than $\pi/2$. To put that another way, the closed loop output delay

t_o/g_{oif} cannot be reduced below $2t_i/\pi$. This is in line with our intuition that a process cannot be controlled on a time scale shorter than the transport delay around the control loop.

Just below the critical value $\pi/2$, the response to a step function in the reference or disturbance is a series of exponentially decreasing oscillations converging to the steady state value, and when $\gamma \leq 1/e$, a non-oscillating exponential convergence to the steady state.

We shall now mathematically demonstrate these results. Differential equations with transport delays can be difficult to solve analytically, so we shall proceed by guessing some general forms of solutions, and proving that these guesses satisfy the equations. A rigorous proof that all solutions are linear combinations of these ones is beyond the scope of this appendix.

We are primarily interested in the behaviour of the control loop when r and d are zero for all time. A trivial solution is for all the variables to also be constantly zero, but of more interest is the existence of solutions which grow without bound (indicating that the loop is unstable), or a demonstration that all solutions decay to zero with time (indicating that the loop is stable).

With zero r and d, we can eliminate all the quantities but o from the equations (1), and obtain

$$\dot{o} + k_o o + k' o_{t-t_i} = 0 \qquad (2)$$

where $k' = k_o g_{oif}$. We could solve the equations for any other of the variables, but for the present purpose it does not matter which: if one is unstable, all the loop variables are unstable.

Given the observed range of behaviours exhibited by the system in numerical simulation, we guess that there will exist solutions of the following form, for suitably chosen constants a, b, ω, and α:

$$o = ae^{bt/t_i} \cos(\omega t/t_i + \alpha)$$

(We have introduced the scaling of t by t_i for later convenience.) Since the system is time-invariant, α will be undetermined: for any solution of this form, the function with any other value of α must also be a solution. So we can take $\alpha = 0$. Because the system is linear, a will also be undetermined: any multiple of a solution is also a solution. We can therefore take $a = 1$. The interesting quantities are then b and ω. b will be positive for an increasing function, and negative for a decreasing function. When ω is zero, we have a pure exponential, while for other values it is an exponential oscillation.

Substituting our trial solution into equation (2), dividing by the common factor ae^{bt}, and rearranging the terms gives the following equations, where

$C = \cos(\omega t/t_i)$ and $S = \sin(\omega t/t_i)$:

$$C(b/t_i + k_o + k'e^{-b}\cos(\omega)) + S(-\omega/t_i + k'e^{-b}\sin(\omega)) = 0 \quad (3)$$

This will be true for all t if and only if the multipliers of C and S are both zero. These conditions can be manipulated into the following form, where we define $N = t_o/t_i$:

$$e^{-b}\cos(\omega) = -(b + 1/N)/\gamma$$
$$e^{-b}\sin(\omega) = \omega/\gamma$$

These equations can be further rearranged into a form which is geometrically illustrated in Figure 2:

$$\omega^2 + (b + 1/N)^2 = \gamma^2 e^{-2b} \quad (4)$$
$$\cos\omega = -(b + 1/N)/(\gamma e^{-b}) \quad (5)$$

Recall that γ and N are given parameters of the system, while b and ω are values to be found. The depiction of ω as an exterior angle of the triangle is

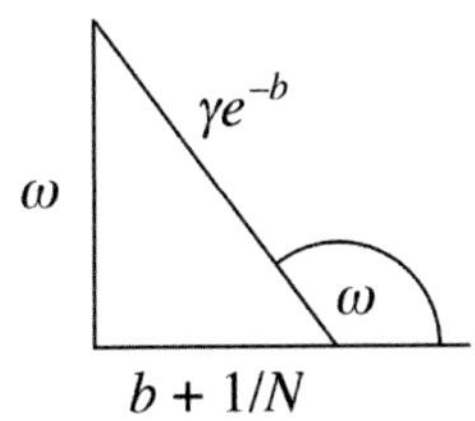

Figure 2: Relations between γ, N, b and ω

slightly misleading. If the depicted angle is ω_0, then $2n\pi \pm \omega_0$ is also a solution to equation (5) for every integer n.

There is no simple formula to give b and ω in terms of γ and N. To proceed further we shall examine the special cases which separate the different types of solution, and plot some illustrative examples.

1.3.1 Static oscillations

For a steady state oscillation, neither growing nor decaying, b must be zero. We can then obtain ω explicitly from equation (4):

$$\omega = \gamma\sqrt{1 - 1/(\gamma N)^2}$$

Equation (5) can then be read as imposing a relation between γ and N.

γN is the loop gain, and when this is large, ω is approximately γ. Since ω is also the indicated angle in the diagram (or $2n\pi \pm$ that angle for any n), this dictates a relationship between γ and N that must hold for a steady state oscillation to be possible.

When N is large, the smallest possible value of ω is approximately $\pi/2$, and γ has approximately the same value. This means that the steady state oscillation has a period of about $4t_i$. Higher values of γ allow faster oscillations, with periods of approximately $4t_i/(2n+1)$ for integers n. However, these are only of theoretical interest, as larger values of γ also allow unstable increasing oscillations.

1.3.2 Exponential non-oscillating solutions

For an exponential solution without oscillations, ω must be zero. Equation (3) then gives us

$$e^{-b} + (b + 1/N)/\gamma \quad = \quad 0 \tag{6}$$

Since the first term is always positive, this equality can only be satisfied for negative values of b. Thus the non-oscillating exponential solutions are all decaying ones.

For some values of γ and N, there are no solutions. For other values, there are two solutions, and on the boundary there is one. The graphs of e^{-b} and $-(b + 1/N)/\gamma$ are illustrated in Figure 3 for $N = 10$ and $\gamma = 0.3$. As γ increases, the straight line becomes flatter; as the loop gain $g_{oif} = \gamma N$ increases, its intercept on the vertical axis approaches zero. The boundary

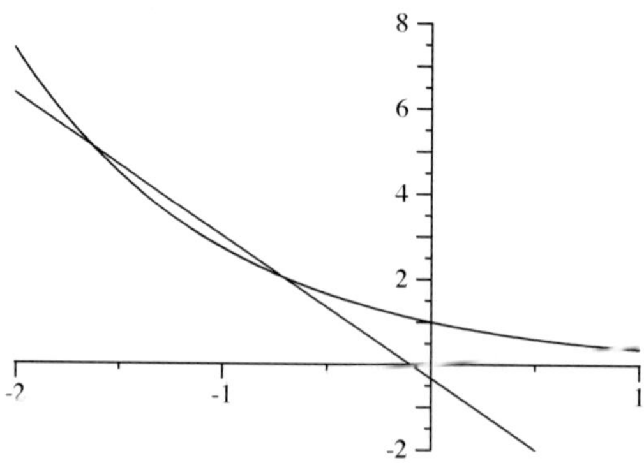

Figure 3: e^{-b} and $-(b + 1/N)/\gamma$. $N = 10$, $\gamma = 0.3$.

value, when the straight line is tangent to the exponential, can be calculated by requiring that in addition, the derivative of the left hand side is zero:

$$-e^{-b} + 1/\gamma \quad = \quad 0 \tag{7}$$

Eliminating e^{-b} from equations (6) and (7) gives us $b = -1 - 1/N$ and $\gamma = e^b$. For large N, b tends to -1, and γ tends to $1/e$. Smaller values of γ allow two exponentials of different time constant, one tending to infinity and the other tending to zero as γ tends to zero. Figure 4 plots γ (vertical axis) against b (horizontal axis). For γ in the range $1/e$ to $\pi/2$ (approximately, for large N),

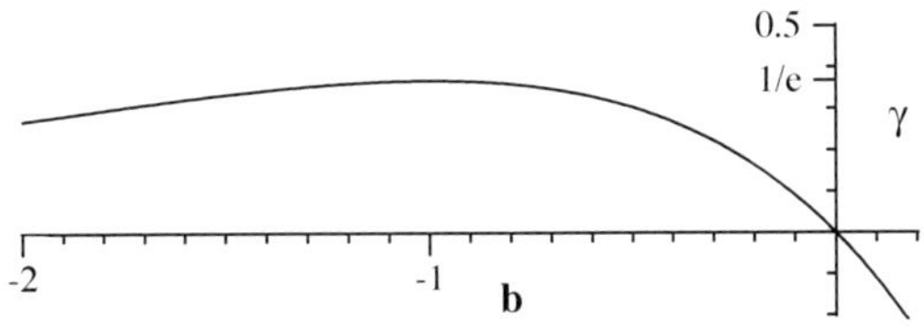

Figure 4: γ and b for non-oscillating solutions for large N

we will see that damped oscillations build up. At $\gamma = 1/e$, the response of the control loop is as fast as possible. The time constant of the loop is then t_i.

1.3.3 Critical damping

From knowledge of the behaviour of damped simple harmonic motion, we may suspect that when γ and N allow exactly one decaying exponential, there is an additional solution of the form $o = ate^{bt/t_i}$. We shall demonstrate this, considering the more general test solution:

$$o \quad = \quad ate^{bt/t_i}\cos(\omega t/t_i + \alpha)$$

As before, we can take $a = 1$ and $\alpha = 0$. If we insert this function into equation (2), we obtain:

$$\begin{array}{c} C(1 - \gamma e^{-b}\cos\omega) + S(\gamma e^{-b}\sin\omega) + \\ Ct(b/t_i + k_o + k' e^{-b}\cos\omega) + St(-\omega/t_i + k' e^{-b}\sin\omega) \end{array} \quad = \quad 0$$

For this to be true for all t, the coefficients of all of C, S, Ct, and St must be zero. This happens if and only if $\omega = 0$, $b = -1 - 1/N$, and $\gamma = e^b$, the values found in the previous section. So the general solution in this case is $o = (a_1 + a_2 t)e^{bt/t_i}$ for any constants a_1 and a_2.

1.3.4 *Oscillating solutions*

We shall now assume that N is large and consider the existence of oscillating solutions. Equations (4) and (5) are approximated by:

$$\begin{aligned} \omega^2 + b^2 &= \gamma^2 e^{-2b} \\ \cos\omega &= -b/(\gamma e^{-b}) \end{aligned} \tag{8}$$

Numerical solution for b in terms of γ yields Figure 5, in which the dashed line is the previous graph of the non-oscillatory solutions. As γ increases from

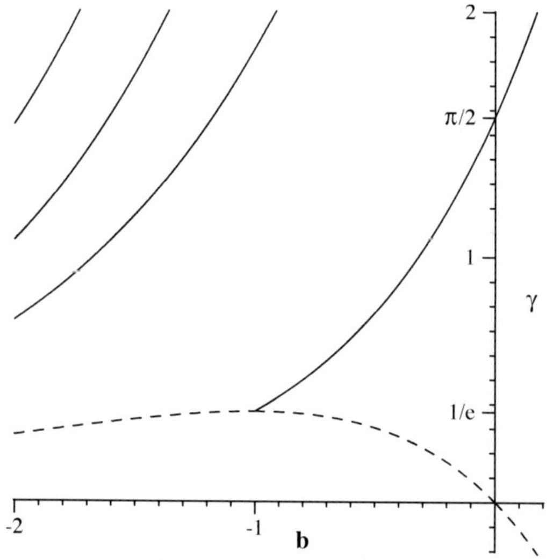

Figure 5: γ and b for oscillating solutions

$1/e$ to $\pi/2$, b increases from -1 to 0. Thus the time constant of the decay of the oscillations becomes longer and longer. For γ above $\pi/2$, b becomes positive, and the system is unstable. The period of the oscillations decreases from infinity at critical damping to $4t_i$ for sustained oscillations, approaching $2t_i$ as γ increases above that point.

The extra branches in the top left of the diagram correspond to the higher frequency solutions of equations (8). For a given value of γ, the behaviour is dominated by the rightmost solution for b.

Figure 6 shows five solutions for the output of the system for various values of γ, 0.1 (overdamped), $1/e$ (critically damped), 1 (underdamped), $\pi/2$ (the edge of stability), and 2 (unstable). We have taken $g_i = g_f = 1$. The horizontal axis is in units of t_i. A plot of the perception looks the same, delayed by t_i. For comparison with the solutions studied in the next section, we have taken the constant reference value to be 1, and chosen solutions which take the value zero when $t = 0$.

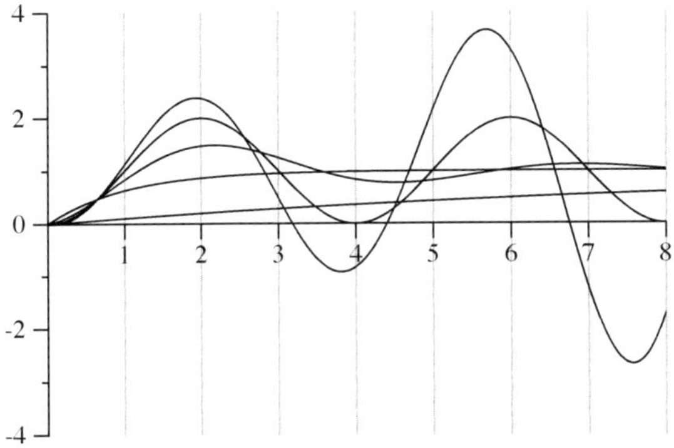

Figure 6: Solutions for o for various values of γ, and $r = 1$.

1.4 Response to a step function

The solutions derived so far have been for zero reference and disturbance. An important aspect of the behaviour of a control system is its response to a step change. We shall give a brief account for the case where N is large.

Assume that all the variables are zero for negative t, and that at $t = 0$ the reference changes to 1 and remains there. During the time interval from 0 to t_i, the input will remain at zero, due to the transport delay. The error will therefore be equal to 1 and (because N is large) the output will ramp linearly upwards over this interval: $o = \gamma t/t_i$.

During the next time interval, from t_i to $2t_i$, the input will be a linearly increasing ramp, the error will be a linearly descending ramp, and the output will be the following quadratic function: $o = \gamma t/t_i - \frac{1}{2}\gamma^2(t/t_i - 1)^2$. When $t = 2t_i$, this quadratic function will begin to appear at the input, and the output from $2t_i$ to $3t_i$ will be the cubic function $o = \gamma t/t_i - \frac{1}{2}\gamma^2(t/t_i - 1)^2 + \frac{1}{6}\gamma^3(t/t_i - 2)^3$. During the nth interval, o will be an nth order polynomial joining smoothly onto the previous interval. This piecewise polynomial solution exhibits the same convergent or divergent behaviour as the exponential-trigonometric solutions derived earlier, depending on the value of γ, and it becomes indistinguishable by eye from those solutions after only a few multiples of t_i. Examples of the output are plotted in Figure 7 for the same values of γ as in Figure 6. As before, $g_i = g_f = 1$, so the perception is a delayed version of the output.

The response to a step change in the disturbance is the same, except that in the stable case the perception settles back to zero.

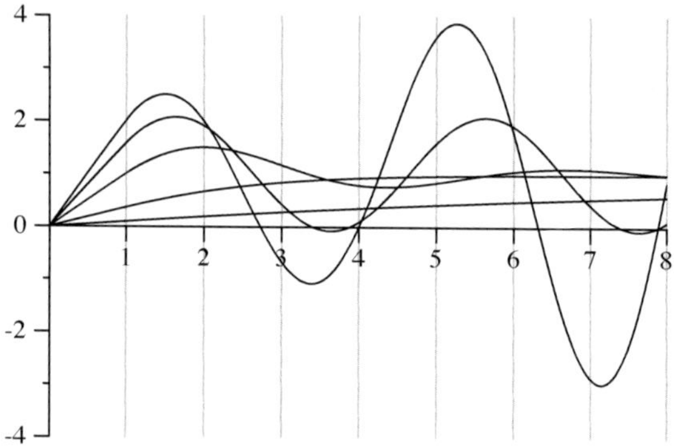

Figure 7: Piecewise polynomial response to a step change in the reference.

1.5 Saturation of input and output

In a real control system, there is always a limit on the amount of output its actuator can produce, and the amount of input its sensor can sense. To model this, we replace the equations $f = g_f o$ and $i = d + f$ by $f = g_f \sigma(o, m_o)$ and $i = \sigma(m_i, d + f)$, where m_o and m_i are the maximum output and input possible, and σ is a function such that when $|x|$ is much less than m, $\sigma(x, m)$ is approximately equal to x, but for x of larger magnitude, $\sigma(x, m)$ is limited to the range $\pm m$. σ might be defined by "hard clipping": $\sigma(x, m)$ is x if $|x| <= m$, m if $x > m$, or $-m$ is $x < -m$, or by a smoother transition, such as $\sigma(x, m) = m \tanh(x/m)$, which makes an exponential approach to the limit.

Input and output clipping have little effect on the system for stable values of γ. In particular, they do not affect its stability. For unstable values, either input or output clipping has the effect of limiting the magnitude of the oscillations.

2 THE TWO-LEVEL CASCADE

Chapter 5 presented a two-level cascade control system, in which the output of the upper controller was the reference for the lower controller. To analyse this, we shall label the reference, perception, error, output, and gain of the upper system by r_0, p_0, ϵ_0, o_0, and k_0. For the lower system the quantities are r_1, p_1, etc. To keep the analysis simple, the controllers are assumed to be proportional controllers, and the environment is assumed to contain no transport delays.

Here are the equations describing the whole system. For the controllers, we have, for $j = 0, 1$:

$$\begin{aligned} \epsilon_j &= r_j - p_j \\ o_j &= k_j \epsilon_j \\ r_1 &= o_0 \end{aligned}$$

For the environment, containing the mass on a damped spring:

$$\ddot{p}_0 = (f - k_d \dot{p}_0 - k_p p_0)/m$$

The connections between the controllers and the environment are these:

$$\begin{aligned} p_1 &= \dot{p}_0 \\ f &= o_1 + d \end{aligned}$$

d is an additional disturbing force. At this point it simplifies the analysis to take the Laplace transform of these equations. Letting $\mathcal{E}$ be the transform of ϵ and similarly for the other variables, the differential equations are transformed into the following algebraic equations.

$$\begin{aligned} \mathcal{E}_j &= \mathcal{R}_j - \mathcal{P}_j & \qquad \mathcal{P}_1 &= s\mathcal{P} \\ \mathcal{O}_j &= k_j \mathcal{E}_j & \mathcal{F} &= \mathcal{O}_1 + \mathcal{D} \\ \mathcal{R}_1 &= \mathcal{O}_0 & s^2\mathcal{P} &= (\mathcal{F} - k_d s\mathcal{P} - k_p \mathcal{P})/m \end{aligned}$$

We can eliminate most of the variables and obtain the following relation between $\mathcal{P}_0$, $\mathcal{R}_0$, and $\mathcal{D}$. We abbreviate $k_0 k_1$ to k_{01}.

$$\mathcal{P}_0 = (\mathcal{R}_0 k_{01} + \mathcal{D})/q(s)$$

where $q(s)$ is the polynomial $ms^2 + s(k_1 + k_d) + k_{01} + k_p$. The steady-state behaviour is found by setting $s = 0$, giving $p_0 = r + d/k_{01}$. So if the system is stable and the gains are large, p_0 tracks r_0, and is disturbed only slightly by d. Whether the system is stable depends on whether the roots of $q(s)$ lie in the left half of the complex plane, and the further to the left they are, the faster the perception tracks the reference and rejects disturbances. These roots are

$$s = \frac{1}{2m}\left(-(k_1 + k_d) \pm \sqrt{(k_1 + k_d)^2 - 4m(k_{01} + k_p)}\right)$$

Since all of the constants are positive, the real part of this is negative. Critical damping is obtained when the roots are equal, i.e. when $(k_1 + k_d)^2 = 4m(k_{01} + k_p)$. The time constant is $2m/(k_1 + k_d)$. When the environmental system has zero damping and spring constant, this is equivalent to $k_1 = 4mk_0$, with a time constant of $2m/k_1 = 1/2k_0$.

For a constant reference r_0, the steady state error is $k_p r_0/(k_{01} + k_p)$. Both the steady state error and the time constant can be made arbitrarily small by choosing sufficiently high k_0 and k_1. As for the basic proportional controller, if the system being controlled contains transport delays, this places limits on the gain and the time constant for the system to be stable. Unlike the case of a single controller, bounds on inputs and outputs can affect stability of the cascade system.

The phenomenon of "non-adaptive adaptation" described in Chapter 5 can be mathematically explored by determining the transfer function between the two variables related in the Bode plots of Figure 5-6. These are the error ϵ_0 and the force output f. By a similar derivation to the one above, we find that, omitting the disturbance:

$$\mathcal{F} \quad = \quad \mathcal{E}_0\, k_{01} n(s)/(n(s) + k_1 s)$$

where $n(s) = ms^2 + sk_d + k_p$. $n(s)$ depends only on the environment, while k_0 and k_1 characterise the control system. As $n(s)$ varies, so the dependency of f on ϵ_0 varies. The frequency and phase plots that are plotted by numerical simulation in Figure 5-6 can be computed by standard methods directly from the transfer function $k_{01}n(s)/(n(s)+k_1 s)$. Figure 8 shows the resulting plots, using the same values of the parameters as are used in Figure 5-6.

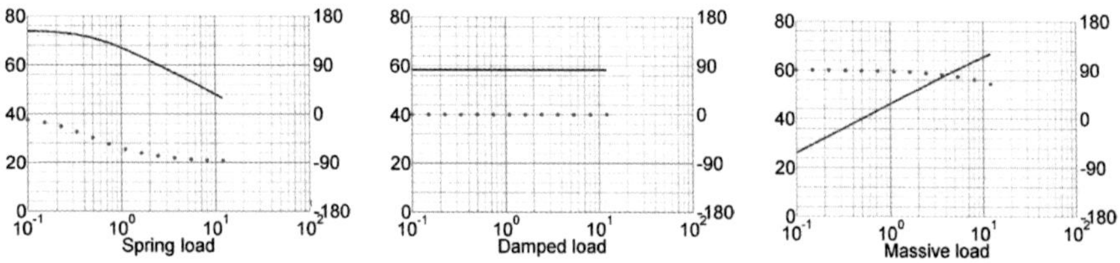

Figure 8: Bode plots relating output force to error. Horizontal axis = frequency (radians/sec.), left axis = gain (decibels, solid line), right axis = phase (degrees, dotted line).

Higher levels of cascades can be similarly analysed, although the degree of the polynomials increases to the point where numerical methods are necessary to determine the location of the roots. For example, the inverted pendulum discussed in Chapter 9 and illustrated on the left of Figure 9 can be controlled by means of a force applied to the cart, using the four-level cascade shown in Figure 10. The pendulum joint is assumed to swing freely. The individual controllers, from the top down, sense the horizontal position of the bob, its velocity, the horizontal offset between bob and cart, and the cart's velocity. It is assumed that the pendulum is initially near-vertical: this controller does not have the capability to swing the pendulum up if it is hanging down. Suitable values for the gains of the four controllers can be found by mathematical

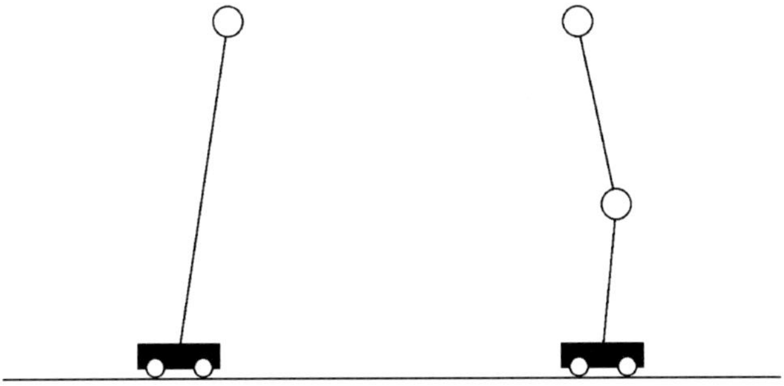

Figure 9: The single and double inverted pendulum

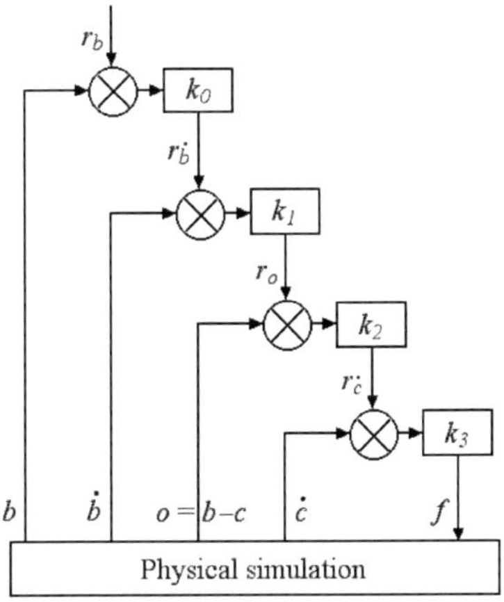

Figure 10: Control hierarchy for the single inverted pendulum

analysis, but it is possible to obtain a system that succeeds in controlling the pendulum by choosing physically reasonable values.

The same technique can even be extended to the double inverted pendulum, for which both joints swing freely and the actuator applies a horizontal force to the cart. This can be similarly controlled by a six-level cascade. Choosing the gains carefully becomes more important for higher order systems. We saw that the two-level system without transport delays was stable for all positive gains, but this is not true for cascades of three or more.

If the single inverted pendulum is being stably controlled at the reference position, and the reference is suddenly moved to the right, the initial action of the system will be to move the cart to the left. Any reasonable method must do this, since that is the only way to make the bob accelerate to the right. As the bob approaches the reference point, the cart will slightly lead the bob, to decelerate it. At least, that is the behaviour that will be observed. It is important to note, however, that the controller does not operate by means of any such plan of action.

3 MULTIDIMENSIONAL CONTROL

Chapter 6 described a system of two controllers, in which each controller controlled a different weighted sum of the same two environmental variables, and the rates of change of these variables were weighted sums of the outputs of the controllers. The equations were:

$$\begin{aligned} P_1 &= a_{11}V_1 + a_{12}V_2 & \qquad V_1 &= k_{11}A_1 + k_{21}A_2 \\ P_2 &= a_{21}V_1 + a_{22}V_2 & \qquad V_2 &= k_{12}A_1 + k_{22}A_2 \end{aligned}$$

Each A_i is the integral of the corresponding error signal. We shall sketch the underlying mathematics of a system of n such controllers and n variables, for any n.

The perceptions of the controllers form an n-element vector P, each element of which is a weighted sum of the components of an n-element vector V describing the environment. That is, $P = AV$ for some matrix A. The output weights are similarly given by a matrix B. Letting R, E, D, and O be the vectors of references, errors, disturbances, and outputs, we have:

$$\begin{aligned} E &= R - P \\ \dot{V} &= BE \\ P &= AV + D \end{aligned}$$

This is a multivariate version of equations (1), simplified by omitting the delay constants and the output damping. Eliminating everything but P, R, and D, we obtain:

$$\dot{P} = AB(R - P) + \dot{D}$$

In the Laplace domain we have $s\mathcal{P} = AB\mathcal{R} - AB\mathcal{P} + s\mathcal{D}$, and so $\mathcal{P} = (AB + Is)^{-1}(AB\mathcal{R} + s\mathcal{D})$. This describes a stable response of the perceptions to the references and disturbances if and only if all the eigenvalues of AB are in the positive half-plane. If we suppose that the perceptual function A is given, and we are free to choose B, then an obvious choice is $B = kA^{-1}$ for any constant k. This gives $\mathcal{P} = \mathcal{R}/(s + k)$, which means that every perceptual variable approaches its reference with a time constant of k, independently of every other perception. Each output has been designed to act only on the corresponding input.

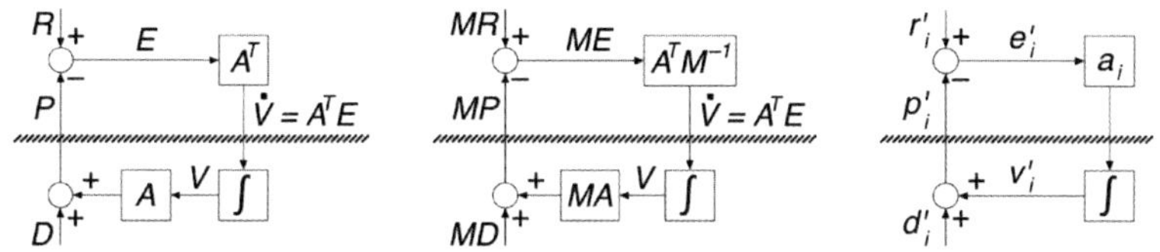

Figure 11: Separation of variables in a multidimensional control system

If A has no inverse, this means that the perceptual variables are not independent. They cannot be simultaneously controlled at independent reference values.

In practice, when the number of variables is large, it may be impractical to compute the inverse of A. An interesting choice of B is the transpose of A (or any multiple of it). (For the example of Chapter 6, this is the same as setting each k_{ij} to be a_{ij}.) When A has an inverse, AA^T is a symmetric positive-definite matrix, that is, its eigenvalues are all real and positive. This implies that taking $B = kA^T$ always gives a stable system. The properties of AA^T imply that there is a matrix M such that MAA^TM^{-1} is diagonal, with all its diagonal elements positive. This means that each of the components of MP is being controlled independently of the others, just as when $B = A^{-1}$, all of the original input variables P are being independently controlled. Figure 11 illustrates this reanalysis of the system as a controller for MP. On the left is the physical organisation of the controller and the environment, the middle shows a mathematically equivalent form, and the diagram on the right illustrates a single controller. The perception p'_i is an element of MP, and similarly r'_i, etc. The constant a_i is one of the eigenvalues of AA^T.

We have assumed that the number of controllers is the same as the number of environmental variables to be controlled. If there are more controlled variables than controllers, some degrees of freedom of the system are not being controlled, and will not resist disturbances. If there are fewer variables than controllers, then there will be a conflict among the controllers: they cannot all succeed in bringing their perceptions to their reference values.

4 REORGANIZATION

Chapter 7 introduced the *E. coli* method of reorganization, whereby a multidimensional controller such as the above can automatically discover a suitable matrix B, without having any knowledge of A. In terms of the analysis of section 3, if AB has any negative eigenvalues, the system is unstable and errors will grow without bound. When reorganization adjusts the output weight matrix B, the eigenvalues of AB move in arbitrary directions (since the changes to B are in arbitrary directions). If the most negative eigenvalue becomes more negative, errors will increase faster and reorganizations will be made more frequently. When reorganization hits on a direction to change B which reduces the instability, it will persist in changing the elements of B in that direction. When the eigenvalues all migrate into the positive half-plane, if any of them have very small real components (implying a large time constant for the system to correct some errors), errors will persist on timescales long enough for reorganization to continue. Eventually, the eigenvalues will all become sufficiently positive that the perceptions will all accurately track the references, and reorganization slows to a crawl.

5 MULTI-LEVEL MULTIDIMENSIONAL CONTROL

Chapter 8 presented a multi-level multi-dimensional controller for an arm. We shall briefly analyse a simpler two-level system in which there are two controllers on each level. The references of the lower controllers are weighted sums of the outputs of the higher controllers. All the controllers are proportional ones, and the bottom-level outputs will be forces causing accelerations of the environmental variables.

The system controls the two-legged robot shown in Figure 12. The legs are telescopic, and we assume that the tilt of the body is small enough that the legs can be considered to be vertical. Two variables are to be controlled: the height h and the pitch p of the body.

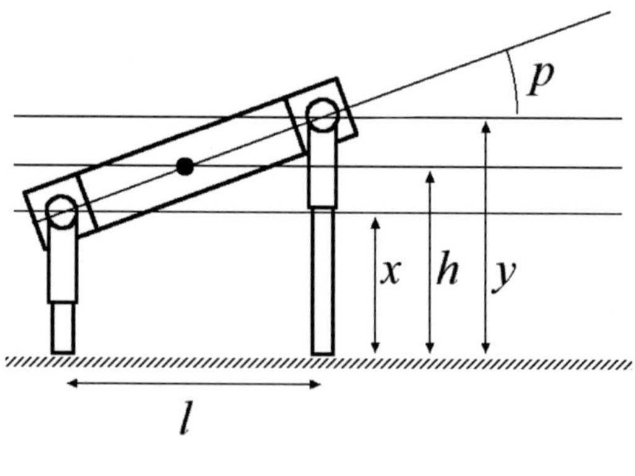

Figure 12: Two-legged robot with one degree of freedom in each leg

The control system is shown in Figure 13. To simplify the analysis, we choose units so that the mass m of the body is 1, and the distance l between the two legs is 2. Having chosen units of mass and length, the moment of inertia of the body will be some value i. There are four proportional controllers. The two upper level controllers control h and p, and the lower level controls $\dot{x}$ and $\dot{y}$. By symmetry it is reasonable to choose $k_x = k_y = k$. Define the following vectors and matrices:

$$H = \begin{pmatrix} h \\ p \end{pmatrix} \quad X = \begin{pmatrix} x \\ y \end{pmatrix} \quad F = \begin{pmatrix} f_x \\ f_y \end{pmatrix}$$

$$R_H = \begin{pmatrix} r_{\dot{x}} \\ r_{\dot{y}} \end{pmatrix} \quad R_{\dot{X}} = \begin{pmatrix} r_h \\ r_p \end{pmatrix} K_H = \begin{pmatrix} k_h & 0 \\ 0 & k_p \end{pmatrix}$$

$$L = \begin{pmatrix} 1 & -1 \\ 1 & 1 \end{pmatrix} \quad A = \begin{pmatrix} \frac{1}{2} & \frac{1}{2} \\ -\frac{1}{2} & \frac{1}{2} \end{pmatrix} \quad C = \begin{pmatrix} 1 & 0 \\ 0 & \frac{1}{i} \end{pmatrix}$$

The equations for the system are then:

$$\begin{aligned} \text{upper layer and linkage:} \quad R_{\dot{X}} &= LK_H(R_H - H) \\ \text{lower layer:} \quad F &= k(R_{\dot{X}} - \dot{X}) \\ \text{environment:} \quad \ddot{H} &= CAF \\ \text{robot geometry:} \quad H &= AX \end{aligned}$$

Eliminating X and noting that $A = L^{-1}$, we solve for H:

$$\ddot{H} + kC\dot{H} + 2kCK_H H = 2kCK_H R$$

Since C and K_H are diagonal, this vector equation separates into its components as:

$$\ddot{h} + k\dot{h} + kk_h h = kk_h r_h \qquad \ddot{p} + k\dot{p}/i + kk_p p/i = 2kk_p r_p/i$$

This describes damped harmonic motion for each variable separately. Critical damping is obtained when $k = 4k_h = 4k_p/i$.

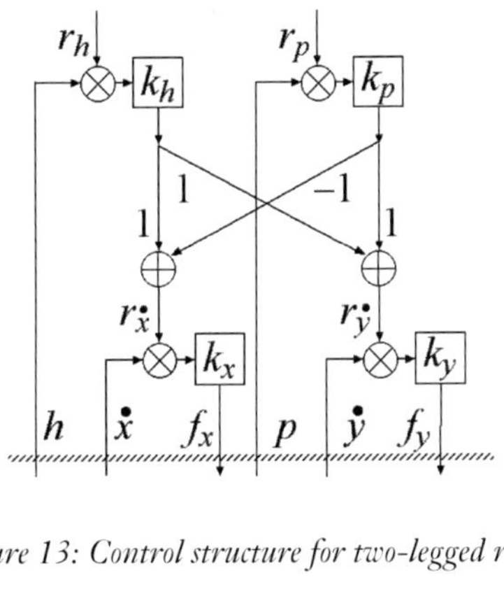

Figure 13: Control structure for two-legged robot

It is instructive to consider what happens if we change the linkage matrix L. Suppose that on the connection from the h controller to the $\dot{x}$ controller, we replace 1 by some number α. L becomes the matrix $\left(\begin{smallmatrix} \alpha & -1 \\ 1 & 1 \end{smallmatrix}\right)$. The equations for h and p become:

$$\begin{aligned} \ddot{h} + k\dot{h} + kk_h h(1+\alpha)/2 &= kk_h r_h(1+\alpha)/2 \\ \ddot{p} + k\dot{p}/i + kk_p p/i &= kk_p r_p/i + kk_h(r_h - h)(1-\alpha)/2 \end{aligned}$$

When $\alpha = 1$, this is the original system. For $\alpha = 0$, the height and pitch control interact, but both still reach their reference value. Note that the height controller is then operating only through the y actuator.

As α approaches -1, the response time of the height controller becomes longer and longer. When $\alpha = -1$, the linkage matrix is singular, which means that the height and pitch controllers are attempting to control different variables by means of identical actions, which is not possible. Control over both variables is lost. For $\alpha < -1$, the system is unstable. Examples of the resulting trajectories are displayed in Figure 14. In each graph, the horizontal axis is the height and the vertical is pitch. Eight trajectories are shown in each one, starting from the open circles and ending at the filled circles. The references for height and pitch are zero. The values chosen for the constants are $i = 0.8$, $k = 4$, $k_h = 1$, and $k_p = 1.25$ (which gives critical damping for both height and pitch). In the five graphs, α takes the values 1, 0, -0.7, -1, and -1.2 respectively.

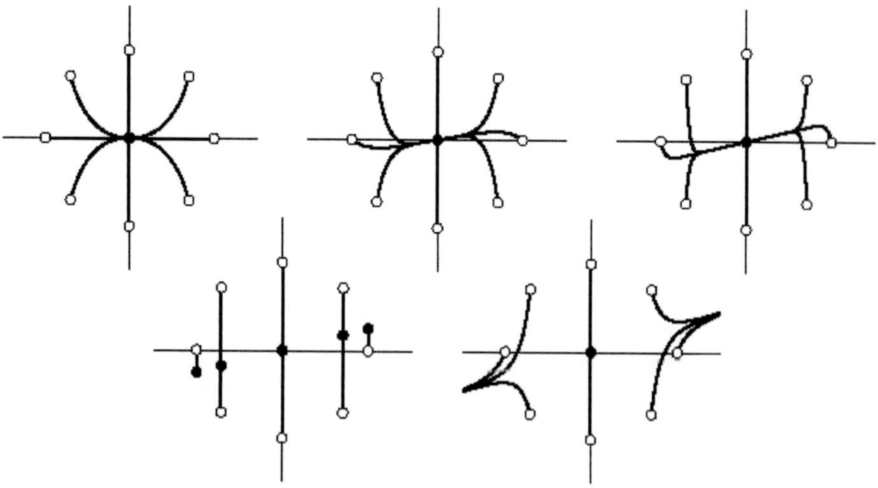

Figure 14: Trajectories of interacting controllers. See text for explanation.

6 A NOTE ON MODEL-BASED CONTROL

Model-based control is a general approach to controller design in which the controller contains a model of the system being controlled. That is, there are physical signals present in the controller (or when implemented in software, program variables), and relationships among them which are designed to imitate the structure of the system being controlled. The purpose of this arrangement is to control the system by calculating what controller outputs, applied to the model, will bring the controlled variable to the desired value.

While model-based control is a major branch of modern engineering control theory, it is important to realise that the fundamental task of control does not inherently require the controller to contain a model of the thing controlled. Neither does success at controlling a variable imply the presence of any model. The designer of a controller may use a model of the system to arrive at a working design, but that model is in the designer's head, not in the controller.

The reader will observe that none of the control systems described in this book contain models. The basic proportional controller contains no model of anything, and neither do the hierarchical networks in the later chapters. Even the adaptive system of Chapter 7 does not achieve control by constructing a model. It constructs a matrix that tends to an approximate inverse of the matrix describing the environment, but a large amount of imprecision still allows successful control of all the variables. Whether any of the control systems that have arisen in the evolution of living organisms make use of models is an interesting question, not answered here. However, the demonstrations in this book show that a large range of functionality can be produced by model-free control systems.

References

[Lei04] J. R. Leigh. *Control Theory: A Guided Tour (2nd ed.)*. Institution of Engineering and Technology, U.K., 2004.

Dr. Richard Kennaway
East Anglia University
Norwich, England UK

INDEX

A

B

C

D

E

F

G

H

I

J

K

L

M

N

O

R

S